Selected Titles in This Series

726 **Victor L. Shapiro,** Singular quasilinearity and higher eigenvalues, 2001

725 **Jean-Pierre Rosay and Edgar Lee Stout,** Strong boundary values, analytic functionals, and nonlinear Paley-Wiener theory, 2001

724 **Lisa Carbone,** Non-uniform lattices on uniform trees, 2001

723 **Deborah M. King and John B. Strantzen,** Maximum entropy of cycles of even period, 2001

722 **Hernán Cendra, Jerrold E. Marsden, and Tudor S. Ratiu,** Lagrangian reduction by stages, 2001

721 **Ingrid C. Bauer,** Surfaces with $K^2 = 7$ and $p_g = 4$, 2001

720 **Palle E. T. Jorgensen,** Ruelle operators: Functions which are harmonic with respect to a transfer operator, 2001

719 **Steve Hofmann and John L. Lewis,** The Dirichlet problem for parabolic operators with singular drift terms, 2001

718 **Bernhard Lani-Wayda,** Wandering solutions of delay equations with sine-like feedback, 2001

717 **Ron Brown,** Frobenius groups and classical maximal orders, 2001

716 **John H. Palmieri,** Stable homotopy over the Steenrod algebra, 2001

715 **W. N. Everitt and L. Markus,** Multi-interval linear ordinary boundary value problems and complex symplectic algebra, 2001

714 **Earl Berkson, Jean Bourgain, and Aleksander Pełczynski,** Canonical Sobolev projections of weak type $(1,1)$, 2001

713 **Dorina Mitrea, Marius Mitrea, and Michael Taylor,** Layer potentials, the Hodge Laplacian, and global boundary problems in nonsmooth Riemannian manifolds, 2001

712 **Raúl E. Curto and Woo Young Lee,** Joint hyponormality of Toeplitz pairs, 2001

711 **V. G. Kac, C. Martinez, and E. Zelmanov,** Graded simple Jordan superalgebras of growth one, 2001

710 **Brian Marcus and Selim Tuncel,** Resolving Markov chains onto Bernoulli shifts via positive polynomials, 2001

709 **B. V. Rajarama Bhat,** Cocylces of CCR flows, 2001

708 **William M. Kantor and Ákos Seress,** Black box classical groups, 2001

707 **Henning Krause,** The spectrum of a module category, 2001

706 **Jonathan Brundan, Richard Dipper, and Alexander Kleshchev,** Quantum Linear groups and representations of $GL_n(\mathbb{F}_q)$, 2001

705 **I. Moerdijk and J. J. C. Vermeulen,** Proper maps of toposes, 2000

704 **Jeff Hooper, Victor Snaith, and Min van Tran,** The second Chinburg conjecture for quaternion fields, 2000

703 **Erik Guentner, Nigel Higson, and Jody Trout,** Equivariant E-theory for C^*-algebras, 2000

702 **Ilijas Farah,** Analytic guotients: Theory of liftings for quotients over analytic ideals on the integers, 2000

701 **Paul Selick and Jie Wu,** On natural coalgebra decompositions of tensor algebras and loop suspensions, 2000

700 **Vicente Cortés,** A new construction of homogeneous quaternionic manifolds and related geometric structures, 2000

699 **Alexander Fel′shtyn,** Dynamical zeta functions, Nielsen theory and Reidemeister torsion, 2000

698 **Andrew R. Kustin,** Complexes associated to two vectors and a rectangular matrix, 2000

697 **Deguang Han and David R. Larson,** Frames, bases and group representations, 2000

(*Continued in the back of this publication*)

Singular Quasilinearity and Higher Eigenvalues

MEMOIRS
of the
American Mathematical Society

Number 726

Singular Quasilinearity and Higher Eigenvalues

Victor L. Shapiro

September 2001 • Volume 153 • Number 726 (second of 5 numbers) • ISSN 0065-9266

American Mathematical Society
Providence, Rhode Island

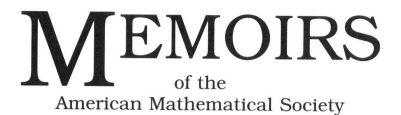

2000 *Mathematics Subject Classification.*
Primary 35J60, 35J70, 35K20, 35K40, 35K55; Secondary 33C50, 42C05, 42C10.

Library of Congress Cataloging-in-Publication Data

Shapiro, Victor L. (Victor Lenard), 1924–
Singular quasilinearity and higher eigenvalues / Victor L. Shapiro.
p. cm. — (Memoirs of the American Mathematical Society, ISSN 0065-9266 ; no. 726)
"September 2001, volume 153, number 726 (second of 5 numbers)."
Includes bibliographical references.
ISBN 0-8218-2717-0
1. Differential equations, Elliptic. 2. Differential equations, Parabolic. 3. Boundary value problems. 4. Singularities (Mathematics). I. Title. II. Series.

QA3 .A57 no. 726
[QA377]
510 s—dc21
[515′.353] 2001032798

Memoirs of the American Mathematical Society

This journal is devoted entirely to research in pure and applied mathematics.

Subscription information. The 2001 subscription begins with volume 149 and consists of six mailings, each containing one or more numbers. Subscription prices for 2001 are $494 list, $395 institutional member. A late charge of 10% of the subscription price will be imposed on orders received from nonmembers after January 1 of the subscription year. Subscribers outside the United States and India must pay a postage surcharge of $31; subscribers in India must pay a postage surcharge of $43. Expedited delivery to destinations in North America $35; elsewhere $130. Each number may be ordered separately; *please specify number* when ordering an individual number. For prices and titles of recently released numbers, see the New Publications sections of the *Notices of the American Mathematical Society.*

Back number information. For back issues see the *AMS Catalog of Publications.*

Subscriptions and orders should be addressed to the American Mathematical Society, P. O. Box 845904, Boston, MA 02284-5904. *All orders must be accompanied by payment.* Other correspondence should be addressed to Box 6248, Providence, RI 02940-6248.

Memoirs of the American Mathematical Society is published bimonthly (each volume consisting usually of more than one number) by the American Mathematical Society at 201 Charles Street, Providence, RI 02904-2294. Periodicals postage paid at Providence, RI. Postmaster: Send address changes to Memoirs, American Mathematical Society, P. O. Box 6248, Providence, RI 02940-6248.

This publication is indexed in *Science Citation Index*®, *SciSearch*®, *Research Alert*®, *CompuMath Citation Index*®, *Current Contents*®/*Physical, Chemical & Earth Sciences.*
Printed in the United States of America.

∞ The paper used in this book is acid-free and falls within the guidelines established to ensure permanence and durability.
Visit the AMS home page at URL: `http://www.ams.org/`

10 9 8 7 6 5 4 3 2 1 06 05 04 03 02 01

To Flo

Contents

ABSTRACT. This research monograph establishes many new results in the area of singular quasilinear elliptic and parabolic partial differential equations and is mainly motivated by three of the author's previous papers, [Sh1], [LfS], and [LgS]. The singularities that arise are described by weights which often can be associated with special functions like the Hermite polynomials and the associated Legendre functions.

Chapter 1 of this monograph deals with a singular quasilinear elliptic operator Q which is a perturbation of a singular elliptic operator L and contains results which improve upon those in [Sh1] in three different directions. The first direction is to obtain results for Q at the higher eigenvalues of L. In particular, Theorem 1 establishes a quasilinear analogue of the Fredholm alternative for Q at specific higher eigenvalues of L; Theorem 4, motivated by [BdF] , gives a double resonance result for Q at all the eigenvalues of L; and Theorem 6 establishes a quasilinear resonance result for Q (see [LaL] and [Wi]) at every L-pseudo-eigenvalue, a new concept.

The second direction of improvement over [Sh1] occurs in Theorems 10-12 where at the first eigenvalue, superlinear extensions of the main results of [Sh1] are obtained via Theorem 9, which constitutes a new Sobolev-type compact imbedding for weighted spaces.

The third direction of improvement is in permitting singularities similar to those which arise in the study of higher order Bessel and associated Legendre functions, neither of which are allowed in [Sh1]. These singularities are covered by the theorems of Chapter 1, and many examples using said functions are given.

Chapter 2 studies time-periodic singular quasilinear parabolic differential equations and extends results previously given in [LfS] and [LgS] and is motivated by semilinear results of the type in [BN] and [CL]. Theorem 1 obtains a double resonance time-periodic result for D_tu-Qu at the higher eigenvalues of L without restriction where Q is an $\#\tilde{H}$-perturbation of L. A number of examples are given to illustrate this new concept.

Theorem 2 of Chapter 2 deals with a time -periodic singular quasilinear reaction-diffusion system at the higher eigenvalues of the singular elliptic operator L. The idea of an $\tilde{H}$-L-pseudo-eigenvalue is introduced and two examples of quasilinear reaction-diffusion systems illustrating this concept are given, one of which is of activator-inhibitor type, an important notion in mathematical biology.

Mathematics Subject Classification 2000. Primary: 35J60, 35J70, 35K20, 35K40, 35K55; Secondary: 33C50, 42C05, 42C10

Key Words and Phrases: Quasilinear elliptic and parabolic partial differential equations, special functions, time periodic, reaction-diffusion systems, superlinear, weighted compact imbedding, double resonance, higher eigenvalues, singular.

Chapter 1

Quasilinear Elliptic Equations

1.1 Introduction

This[1] chapter deals with singular quasilinear elliptic operators which are perturbations of a singular elliptic operator L and contains results which improve upon the author's previous paper [Sh1] in three different ways, namely (i) by working at higher eigenvalues λ_j, (ii) by allowing superlinearities, and (iii) by permitting new types of singularities. To deal with the situation at higher eigenvalues, the new notion of an L-pseudo-eigenvalue is introduced. To handle superlinearities, a new Sobolev-type compact imbedding theorem is proved. The new singularities that are introduced here but that are not found in [Sh1] are those of the type which are related to the higher order Bessel functions and associated Legendre functions.

The results in this chapter were motivated by the author's two previous works on the subject, namely [Sh1] and [Sh2]. These two papers in turn were motivated by the work of Browder [Br], Leray and Lions [LeL], deFigueredo and Gossez [dFG], and Brezis and Nirenberg [BN], and also, Landesman and Lazer [LaL], Williams [Wi], and Kazdan and Warner [KW]. Theorem 4 below was motivated by the work of Berestycki and deFigueredo [BdF] and Robinson [Rob], Theorem 5 by the work of Hetzer [He] and Rumbos [Rum], and Theorem 8 by the author's paper [Sh3].

Let $\Omega \subset R^N, N \geq 1$, be an open (possibly unbounded) set and let

[1]Received by the editor July 20, 1998 and in revised form on December 6, 1999

$\varrho(x), p_i(x) \in C^0(\Omega)$ be positive functions with the property that

$$\int_\Omega \varrho(x)\,dx < \infty \text{ and } \int_\Omega p_i(x)\,dx < \infty \text{ for } i = 1, ..., N. \tag{1.1}$$

Also, let $q(x) \in C^0(\Omega)$ be a nonnegative function and let $\Gamma \subset \partial\Omega$ designate a fixed closed set. (q may be identically zero. Also Γ may be the empty set.) We introduce the pre-Hilbert space:

$$C^1_{p,q,\varrho}(\Omega, \Gamma) = \{u \in C^0(\bar{\Omega}) \cap C^2(\Omega) : u(x) = 0\ \forall x \in \Gamma; \tag{1.2}$$

$$\int_\Omega [\sum_{i=1}^n |D_i u|^2 p_i + u^2 (q + \rho)]\,dx < \infty\},$$

where $p = (p_1, ..., p_n)$ and $D_i u = \partial u / \partial x_i$. In $C^1_{p,q,\varrho}(\Omega, \Gamma)$, we have the inner product

$$< u, v >_{p,q,\varrho} = \int_\Omega \left[\sum_{i=1}^n p_i D_i u D_i v + (q + \varrho) uv\right] dx. \tag{1.3}$$

$H^1_{p,q,\varrho}(\Omega, \Gamma)$ will designate the real Hilbert space that we obtain by completing $C^1_{p,q,\varrho}(\Omega, \Gamma)$ by the method of Cauchy sequences with respect to the norm $\|u\|_{p,q,\varrho} =< u, u >^{\frac{1}{2}}_{p,q,\varrho}$. L^2_ϱ will be the real Hilbert space with the inner product $< u, v >_\varrho = \int_\Omega uv\varrho dx$. In a similar manner we have the spaces $L^2_{p_i}$, $i = 1, ..., N$, and L^2_q. Hence we see from (1.3) that

$$< u, v >_{p,q,\varrho} = \sum_{i=1}^N < D_i u, D_i v >_{p_i} + < u, v >_\varrho + < u, v >_q . \tag{1.4}$$

The last term in (1.4) will be taken to be zero in case q = 0. Also, in the sequel, sometimes we shall write $H^1_{p,q,\varrho}$ for $H^1_{p,q,\varrho}(\Omega, \Gamma)$ and $C^1_{p,q,\varrho}$ for $C^1_{p,q,\varrho}(\Omega, \Gamma)$.

Next, we assume

$$\begin{aligned} &(i)\ a_0, a_{ij} \in C^0(\Omega) \cap L^\infty(\Omega), i, j = 1, ..., N, \\ &(ii)\ a_{ij}(x) = a_{ji}(x)\ \forall x \in \Omega \text{ and i,j = 1,...,N}, \\ &(iii)\ a_0(x) \geq \varepsilon_0 > 0\ \forall x \in \Omega, \\ &(iv) \sum_{i,j=1}^N a_{ij}(x)\xi_i\xi_j \geq c_0|\xi|^2 \quad \forall x \in \Omega, \xi \in \mathbf{R}^N \text{ where } c_0 > 0, \end{aligned} \tag{1.5}$$

and introduce the (possibly singular) operator

$$Lu = -\sum_{i,j=1}^{N} D_i \left[p_i^{\frac{1}{2}} p_j^{\frac{1}{2}} a_{ij} D_j u \right] + a_0 u q \tag{1.6}$$

and the two-form

$$\mathcal{L}(u,v) = \sum_{i,j=1}^{N} \int_\Omega p_i^{\frac{1}{2}} p_j^{\frac{1}{2}} a_{ij} D_j u D_i v + < a_0 u, v >_q \tag{1.7}$$

for $u, v \in H^1_{p,q,\varrho}$.

We have the possibility of (1.6) being singular because the p_i 's may tend to zero on all or part of $\partial\Omega$, *or* Ω may be unbounded, or both.

We shall say (Ω, Γ) is a $\mathrm{V_L}$ - region if the following two facts obtain:

($\mathrm{V_L}$−1) There exists a complete orthonormal system $\{\varphi_n\}_{n=1}^{\infty}$ in L^2_ϱ. Also $\varphi_n \in H^1_{p,q,\varrho}(\Omega,\Gamma) \cap C^2(\Omega)\ \forall n$.

($\mathrm{V_L}$−2) There exists a sequence of eigenvalues $\{\lambda_n\}_{n=1}^{\infty}$with $0 \le \lambda_1 < \lambda_2 \le \lambda_3 \le ... \le \lambda_n \to \infty$ such that $\mathcal{L}(\varphi_n, v) = \lambda_n \langle \varphi_n, v \rangle_\varrho\ \forall v \in H^1_{p,q,\varrho}(\Omega,\Gamma)$. Also $\varphi_1 > 0$ in Ω.

Now there are many examples using special functions to illustrate $\mathrm{V_L}$ - regions. In particular choosing $q = 0$ in (1.7), $a_{ij} = \delta_{ij}$ (the Kronecker -δ) and $p_1 = ... = p_n$, we shall show in §3 that the six examples given in [Sh1, pp. 1413 - 1415] give rise to $\mathrm{V_L}$ - regions. However $q \neq 0$ enables us to deal with associated Legendre functions and higher order Bessel functions. So we shall illustrate a number of further examples in §3 that give rise to $\mathrm{V_L}$ - regions that are not covered by the condition $O_{p,\varrho}(\Omega)$ given in [Sh1, p.1413] .

We also remark at this juncture that for appropriate p,q, and ϱ,the concept of a V_L−region includes that of the familiar Neumann condition for the the eigenfunctions φ_n if they are in $C^2(\bar{\Omega})$ and Ω has a smooth boundary. To see this say for L=Laplace operator, choose p_i=1, for $i = 1, ..., n$ and $q = a_0 = 0$ in (1.6). Also, take Γ =empty set with $\varrho = 1$ in the definition of $H^1_{p,q,\varrho}(\Omega,\Gamma)$ given just below (1.3). Then $(V_L - 2)$ becomes

$$\int_\Omega \nabla\varphi_n \cdot \nabla v = \lambda_n \int_\Omega \varphi_n v \text{ for } v \in W^{1,2}(\Omega)$$

where $W^{1,2}(\Omega)$ is the familiar Sobolev space, [GT, p. 153]. An integration by parts (i.e., an application of Green's formula) to the left-hand side of this last equation leads to

$$\int_{\partial\Omega} v\frac{\partial\varphi_n}{\partial\nu} = 0 \text{ for } v \in W^{1,2}(\Omega)$$

where ν is the outer unit normal and hence (see [Ke, p.124]) to $\frac{\partial\varphi_n}{\partial\nu} = 0$ on $\partial\Omega$, which is the Neumann condition.

Next we introduce two different quasilinear operators. The first is the operator

$$Qu = -\sum_{i=1}^{N} D_i[p_i^{\frac{1}{2}} A_i(x, u, Du)] + qB_0(x)u. \tag{1.8}$$

We make the following assumptions concerning A_i $(i = 1, ..., N)$ and B_0 :

(Q-1) $A_i(x, s, \xi) : \Omega \times \mathbf{R} \times \mathbf{R}^N \rightarrow \mathbf{R}$ and satisfies the Caratheodory conditions (i.e., $A_i(x, s, \xi)$ is measurable for x in Ω for every fixed (s, ξ) $\in$ $\mathbf{R} \times \mathbf{R}^N$ and is continuous in (s,ξ) for a.e. $x \in \Omega$);

(Q-2) There exists $h_i^* \geq 0$ with $h_i^* \in L^2_{p_i}$,$i = 1, \ldots, N$,and a positive constant c_1 such that for a.e. $x \in \Omega$,

$$| A_i(x, s, \xi) | \leq c_1 \sum_{j=1}^{N} p_j^{\frac{1}{2}} (|s| + |\xi_j| + |h_j^*|)$$

or

$$| A_i(x, s, \xi) | \leq c_1 \sum_{j=1}^{N} p_j^{\frac{1}{2}} (|\xi_j| + |h_j^*|)$$

accordingly as L^2_ϱ is or is not continuously imbedded in every $L^2_{p_i}$ (i=1,...,N);

(Q-3) There exists a positive constant c_2 such that

$$\sum_{i=1}^{N} p_i^{\frac{1}{2}} (x) A_i(x, s, \xi)\xi_i \geq c_2 \sum_{i=1}^{N} p_i (x) \xi_i^2$$

for a.e. $x \in \Omega$ and $\forall (s, \xi) \in \mathbf{R} \times \mathbf{R}^N$;

(Q-4)

$$\sum_{i=1}^{N} p_i^{\frac{1}{2}} (x) [A_i (x, s, \xi) - A_i(x, s, \xi')] (\xi_i - \xi_i') > 0 \text{ for a.e. } x \in \Omega,$$

$$\forall s \in \mathbf{R} \text{ ,and } \forall \xi,\xi' \in \mathbf{R}^N \text{ with } \xi \neq \xi';$$

(Q-5) $B_0(x) \in C^0(\Omega) \cap L^\infty(\Omega)$ is a nonnegative function.

The second quasilinear operator that we introduce will be involved with $\sigma(u) = [\sigma_0(u), \sigma_1(u), ..., \sigma_N(u)]$ where σ_i for i = 0, 1..., N meets the first three of the following four conditions:

$(\sigma - 1)$ $\sigma_i : H^1_{p,q,\varrho} \to \mathbf{R}$ is weakly sequentially continuous;
$(\sigma - 2)$ $\exists \eta_0 > 0$ and η_1 *s.t.* $\eta_0 \leq \sigma_i(u) \leq \eta_1$ $\forall u \in H^1_{p,q,\varrho}$;
$(\sigma - 3)$ $lim_{\|u\|_\varrho \to \infty} \sigma_i(u) = 1$ where $\|u\|^2_\varrho = < u, u >_\varrho$;
$(\sigma - 4)$ $\sigma_i(u) \leq 1$ $\forall u \in H^1_{p,q,\varrho}$.

We set, using (1.5),

$$Mu = -\sum_{i,j=1}^{N} D_i \left[p_i^{\frac{1}{2}} p_j^{\frac{1}{2}} \sigma_i^{\frac{1}{2}}(u) \sigma_j^{\frac{1}{2}}(u) a_{ij} D_j u \right] + a_0 \sigma_0 q u \tag{1.9}$$

where we assume for Mu, $(\sigma - 1) - (\sigma - 3)$ but not necessarily $(\sigma - 4)$. $(\sigma - 4)$ will be for future reference.

We return to Qu defined in (1.8) and introduce the two-form

$$\mathcal{Q}(u, v) = \int_\Omega [\sum_{i=1}^{N} p_i^{\frac{1}{2}} A_i(x, u, Du) D_i v] dx + < B_0 u, v >_q \tag{1.10}$$

$\forall u, v \in H^1_{p,q,\varrho}$. We see from (Q-2) and (Q-5) that $\mathcal{Q}(u, v)$ *is* well-defined for u,v $\in H^1_{p,q,\varrho}$. Also it is to be understood that if q = 0, then the last term in (10) does not appear. (Similarly, for (1.7).)

In this chapter, we shall say Q is #-related to L if the following two conditions hold:

$$(i)\ lim_{\|u\|_{p,q,\varrho} \to \infty} [\mathcal{Q}(u, v)\text{-}\mathcal{L}(u, v)] / \|u\|_{p,q,\varrho} = 0 \quad \text{uniformly for } \|v\|_{p,q,\varrho} \leq 1,$$

(ii) if $q(x)$ is not identically zero, $B_0(x) \geq a_0(x)$ in Ω.

(1.11)

Another way to view the asymptotic relationship stated in (i) above is to say

$$\mathcal{Q} - \mathcal{L} = o(\|u\|_{p,q,\varrho}), \text{ uniformly for } \|v\|_{p,q,\varrho} \leq 1 \text{ as } \|u\|_{p,q,\varrho} \to \infty.$$

Also, condition (ii) says the zeroth order term of Q dominates that of L.

In §3, we shall illustrate a number of different quasilinear elliptic operators Q which are #-related to L. Before proceeding we observe that L is #-related to itself. To see this, we have to show that L in (1.6) can be put into the form(1.8). To do this, we set

$A_i(x,s,\xi) = \sum_{j=i}^{N} p_j^{\frac{1}{2}} a_{ij}(x)\xi_j, B_0(x) = a_0(x)$,and

$Qu = -\sum_{i=1}^{N} D_i \left[p_i^{\frac{1}{2}} A_i(x,u,Du) \right] + B_0(x) uq.$

Then an easy check shows that Q meets (Q-1)-(Q-5), that $Qu = Lu$, and $\mathcal{Q}(u,v) = \mathcal{L}(u,v)\ \forall u,v \in H^1_{p,q,\varrho}$. Hence our assertion that L is #-related to itself is established.

1.2 Statement of main results

We intend to obtain various results for Q at the eigenvalues of L under the #-relationship. These results for the most part were motivated by the author's two previous works on the subject, namely [Sh1] and [Sh2]. These two papers in turn were motivated by the work of Browder [Br], Leray and Lions [LeL], deFigueredo and Gossez [dFG], and Brezis and Nirenberg [BN], and also Landesman and Lazer [LaL], Williams [Wi], and Kazdan and Warner [KW]. Theorem 4 below was motivated by the work of Berestycki and deFigueredo [BdF] and Robinson [Ro], and Theorem 5 by Hetzer [He] and Rumbos [Rum].

The first result we obtain will deal with an equation of the following nature

$$Qu = [\ \lambda_{j_0} u + G(u) f\ ]\ \varrho. \tag{2.1}$$

G in (1.12) will meet the following three conditions:

(G-1) G: $H^1_{p,q,\varrho} \to \mathbf{R}$ is weakly sequentially continuous;
(G-2) $\forall \varepsilon > 0$, $\exists c_\varepsilon$ a constant s.t.

$$|G(u)| \le \varepsilon \|u\|_{p,q,\varrho} + c_\varepsilon\ \forall u \in H^1_{p,q,\varrho};$$

(G-3) $G(u) \neq 0\ \forall u \in H^1_{p,q,\varrho}$.

We will also need the notion of an eigenfunction for the nonlinear operator Q. In particular, we say $\phi \in H^1_{p,q,\varrho}$ is a $H^1_{p,q,\varrho}-\ \lambda_{j_0}-$eigenfunction of Q provided

$$\mathcal{Q}(v,\phi) = \lambda_{j_o} < v,\phi >_\varrho \forall v \in H^1_{p,q,\varrho}. \tag{2.2}$$

λ_{j_o} will be called an $H^1_{p,q,\varrho}-$ eigenvalue of Q.

The first theorem we shall establish is the following:

Theorem 1 *Let $\Omega \subset \mathbf{R}^N, N \geq 1$, be an open set, ϱ and $p_i\,(i = 1, ..., N)$ be positive functions in $C^0(\Omega)$ satisfying (1.1), $q \in C^0(\Omega)$ be a nonnegative function, and $\Gamma \subset \partial\Omega$ be a closed set. Let L and Q be given by (1.6) and (1.8) respectively and assume (Q-1) - (Q-5), that Q is #-related to L, and that (Ω, Γ) is a V_L - region. Also, suppose every $\lambda_{j_o}-$ eigenfunction of L is a $H^1_{p,q,\varrho}-$ $\lambda_{j_o}-$ eigenfunction of Q. Suppose furthermore that G meets (G-1) - (G-3) and $f \in L^2_\varrho$. Then a necessary and sufficient condition that $\exists u^* \in H^1_{p,q,\varrho}$ which is a weak solution of (2.1) is that*

$$< f, \varphi >_\varrho = 0 \text{ for every } \lambda_{j_o} - \text{eigenfunction } \varphi \text{ of } L. \qquad (2.3)$$

In §3 of this paper, we shall illustrate several Q's and L's which are #-related and have eigenfunctions in common for certain higher eigenvalues of L. Also we note that the sufficiency of the above theorem is true even if (G-3) does not hold.

It is clear that this theorem constitutes a nonlinear generalization of the familiar Fredholm alternative (see [GT, p. 85] or [BD, p. 649]).

To be quite explicit, what we mean by the statement that u* $\in H^1_{p,q,\varrho}$ is a weak solution of (2.1) is that the following holds:

$$\mathcal{Q}(u^*, v) = \lambda_{j_o} < u^*, v >_\varrho + G(u^*) < f, v >_\varrho \qquad (2.4)$$

$\forall v \in H^1_{p,q,\varrho}$ where $\mathcal{Q}\,(u^*, v)$ is defined by (1.10).

In case q=0 in the above theorem a slightly better result prevails. In order to present this result, we need the following (weaker) version of (G-1) :

$$(G-1)'\, G : H^1_{p,q,\varrho} \rightarrow \mathbf{R} \text{ is continuous in the strong topology.}$$

To be quite explicit, G satisfies $(G-1)'$ means

$$\|u_n - u\|_{p,q,\varrho} \rightarrow 0 \text{ implies } G(u_n) \rightarrow G(u) \text{ as } n \rightarrow \infty.$$

Since every G which satisfies $(G-1)$ also satisfies $(G-1)'$, if we can replace $(G-1)$ by $(G-1)'$ in the above theorem we have a better result.

The result that we establish for this larger class of G is the following :

Corollary 2 *Assume all conditions in the hypothesis of Theorem 1 hold except that G satisfies (G-1)' and (G-2) and that q=0. Suppose furthermore that Ω is a bounded domain. Then a sufficient condition that $\exists u^* \in H^1_{p,\varrho}$ which is a weak solution of (2.1) is that (2.3) holds.*

In the above corollary we have written $H^1_{p,\varrho}$. This Hilbert space is understood to be the same as $H^1_{p,q,\varrho}$ when q=0. Also in the above corollary we do not assume that G meets (G-3). An example of a G that will qualify for Corollary 2 but not in general for Theorem 1 is

$$G(u) = \|u\|^{\frac{1}{2}}_{p,q,\varrho} + 1.$$

In addition, the following remark concerning Corollary 2 will be evident from the proof that we will give for Corollary 2.

Remark 1. *In Corollary 2, in case $p_1 = \cdots = p_n = \varrho$, then the condition that Ω be a bounded domain is not required.*

Next, we return to Mu in (1.9) and introduce the corresponding two-form $\mathcal{M}(u,v) =$

$$\sum_{i,j=1}^{N} \int_\Omega [\, p_i^{\frac{1}{2}} p_j^{\frac{1}{2}} \sigma_i^{\frac{1}{2}}(u)\, \sigma_j^{\frac{1}{2}}(u)\, a_{ij} D_j u D_i v \,]\, dx + \sigma_0(u) < a_0 u, v >_q \qquad (2.5)$$

for u,v $\in H^1_{p,q,\varrho}$ and σ_i meeting (σ - 1) - (σ - 3) for i = 1,...,N.

A similar theorem to Theorem 1 prevails for Mu. In order to present this result we need the following : λ_{j_o} is a near-eigenvalue of M provided λ_{j_o} is an eigenvalue of L and $\exists\, \sigma^\ddagger$ meeting $(\sigma - 1)$ and $(\sigma - 4)$ such that

$$\mathcal{M}(u,\varphi) = \sigma^\ddagger(u)\ \lambda_{j_o} < u, \varphi >_\varrho \forall u \in H^1_{p,q,\varrho} \qquad (2.6)$$

for every λ_{j_o}−eigenfunction φ of L.

We deal with the equation

$$Mu = [\lambda_{j_o} u + G(u) f]\, \varrho, \qquad (2.7)$$

and have the following theorem:

Theorem 3 *Let $\Omega \subset \mathbf{R}^N, N \geq 1$, be an open set, ϱ and $p_i (i = 1,...,N)$ be positive functions in $C^0(\Omega)$ satisfying(1.1), $q \in C^0(\Omega)$ be a nonnegative function, and $\Gamma \subset \Omega$ be a closed set. Let L and M be given by (1.6) and (1.9) respectively and assume $(\sigma-1)-(\sigma-3)$ for M and that (Ω, Γ) is a V_L region. Furthermore, suppose that G meets (G-1)-(G-2), that $f \in L^2_\varrho$, and that λ_{j_o} is a near-eigenvalue of M. Then if (2.3) holds, $\exists\, u^* \in H^1_{p,q,\varrho}$ which is a weak solution of (2.7).*

To be quite explicit, by u* is a weak solution of (2.7), we mean the following:

$$\mathcal{M}(u^*, v) = \lambda_{j_o} < u^*, v >_\varrho + G(u^*) < f, v >_\varrho\ \forall v \in H^1_{p,q,\varrho}. \qquad (2.8)$$

Also, it is easy to see that if N=1, q=0 and σ_1 meets $(\sigma - 4)$, then every eigenvalue of L is a near-eigenvalue of M. Likewise, if $N \geq 1$ and $\sigma_o = \sigma_1 = ... = \sigma_N$ with σ_1 meeting $(\sigma - 4)$, it follows that every eigenvalue of L is a near-eigenvalue of M. In § 3, we give two different examples where the case of equal σ_i 's does not hold but nevertheless M has an infinite number of different near-eigenvalues.

Returning to the differential operator Q, it turns out that as long as Q is #-related to L, we can still get interesting results for Q at a λ_{j_o} -eigenvalue of L even though λ_{j_o} is not an $H^1_{p,q,\varrho}-$ eigenvalue of Q as in Theorem 1. In particular, a double resonance result (see, [*BdF, Theorem* 3] and [*Ro*]) holds for Q. We present this phenomenon in Theorem 4 below but first we need some more notation. We shall be dealing with the equation

$$Qu = [\, \lambda_{j_o} u + f(x,u) \,]\, \varrho \tag{2.9}$$

where λ_{j_o} is an eigenvalue of L of multiplicity j_1. Thus, $\lambda_{j_o+j_1}$ is the next eigenvalue strictly greater than λ_{j_o}. We shall set

$$\gamma = [\lambda_{j_o+j_1} - \lambda_{j_o}]\,/2 \tag{2.10}$$

and make the following assumptions concerning f(x,s): $\Omega \times \mathbf{R} \to \mathbf{R}$.

(f-1) $f(x,s)$ satisfies the usual Caratheodory conditions.

(f-2) $|f(x,s) - \gamma s| \leq \gamma |s| + f_o(x)$ $\forall s \in \mathbf{R}$ and a.e. $x \in \Omega$ where $f_o \in L^2_\varrho$ and γ is defined in (2.10).

Also, we shall set

$$\mathcal{F}^{\pm}(x) = \lim_{s \to \pm\infty} \sup f(x,s)/s,\ \mathcal{F}_{\pm}(x) = \lim_{s \to \pm\infty} \inf f(x,s)/s \tag{2.11}$$

We shall need the following:

$$\int_{\Omega \cap [v>0]} \left[\lambda_{j_o+j_1} - \lambda_{j_o} - \mathcal{F}^{+}\right] v^2 \varrho + \int_{\Omega \cap [v<0]} \left[\lambda_{j_o+j_1} - \lambda_{j_o} - \mathcal{F}^{-}\right] v^2 \varrho > 0 \tag{2.12}$$

for every nontrivial $\lambda_{j_o+j_1}-$eigenfunction v of L;

$$\int_{\Omega \cap [w>0]} \mathcal{F}_{+} w^2 \varrho + \int_{\Omega \cap [w<o]} \mathcal{F}_{-} w^2 \varrho > 0 \tag{2.13}$$

for every nontrivial $\lambda_{j_o}-$ eigenfunction w of L.

The result that we present for the equation (2.9) is the following:

Theorem 4 *Let $\Omega \subset \mathbf{R}^N$, $N \geq 1$, be an open set, ϱ and p_i (i=1,...,N) be positive functions in $C^0(\Omega)$, q be a nonnegative function, and $\Gamma \subset \partial\Omega$ be a closed set. Let L and Q be given by (1.6) and (1.8) respectively and assume (Q-1)-(Q-5), that Q is #-related to L, and that (Ω, Γ) is a V_L-region. Suppose that f satisfies (f-1),(f-2), λ_{j_o} is an eigenvalue of L of multiplicity j_1, and both (2.12) and (2.13) hold. Then $\exists u^* \in H^1_{p,q,\varrho}$ which is a weak solution of (2.9).*

To be quite explicit what we mean by a weak solution of (2.9) is the following:

$$\mathcal{Q}(u^*, v) = \lambda_{j_o} < u^*, v >_\varrho + < f(\cdot,\ u^*), v >_\varrho \qquad (2.14)$$

$\forall v \in H^1_{p,q,\varrho}$. A theorem of the above kind is called a double resonance result because the condition (f-2) allows $\lambda_{j_o} s + f(x,s)$ to take both the value $\lambda_{j_o+j_1} s$ as well as $\lambda_{j_o} s$ for various values of x in Ω. Whenever $\lambda_{j_o} + s^{-1}\ f(x,s)$ interacts with an eigenvalue in a differential equation of the form (2.9) with Q replaced by L, the corresponding theorem is referred to as a resonance-type result (see, [*BN*, Ch IV] and [*La*]). In §3, we shall give examples of a quasilinear Q where the above theorem is false in case (2.12) holds but (2.13) fails to hold.

We will not prove Theorem 4 in Chapter 1; instead we will prove an analogous result for the time-periodic parabolic case in Chapter 2. There should be no difficulty on the reader's part in adjusting the proof given in Chapter 2 to handle the easier case stated here.

We would like to make the following observation concerning Theorem 4. It is an immediate consequence of this theorem that the following interesting fact holds:

Remark 2 *Let $\Omega \subset \mathbf{R}^N, N \geq 1$, be an open set, ϱ and p_i(i=1,...,N) be positive functions in $C^0(\Omega)$, q be a nonnegative function,* and *$\Gamma \subset \partial\Omega$ be a closed set. Let L and Q be given by (1.6) and (1.8) respectively and assume* (Q-1)-(Q-5), *that Q is #-related to L, and that (Ω, Γ) is a V_L-region. Suppose λ_{j_o} is an eigenvalue of L of multiplicity j_1. Then $\exists u^* \in H^1_{p,q,\varrho}$ which is a weak solution of the equation*

$$Qu = [\ \lambda_{j_o} u + cu + g(x,u) + h(x)\ ]\ \varrho$$

where c is a constant with $0 < c < \lambda_{j_o+j_1} - \lambda_{j_o}$, $g(x,s) \in L^\infty(\Omega \times \mathbf{R}) \cap C^0(\Omega \times \mathbf{R})$, and h$\in L^2_\varrho(\Omega) \cap C^0(\Omega)$.

To establish Remark 2, we set

$$f(x,s) = cs + g(x,s) + h(x)$$

and observe that both (f-1) and (f-2) hold. Also an easy computation shows that

$$\mathcal{F}^{\pm}(x) = \mathcal{F}_{\pm}(x) = c$$

for every $x \in \Omega$. Hence both (2.12) and (2.13) hold. So Remark 2 follows immediately from Theorem 4.

In case λ_{j_o} is also an $H^1_{p,q,\varrho}-$ eigenvalue of Q we can obtain resonance results for an f having linear growth which meets both (f-2) and (2.12) but not necessarily (2.13) provided it also meets (f-3),(f-4), and (2.16) below. (We shall give such an example in § 3 which will also be solvable by Theorem 5 stated below.) In order to present a result of this nature we shall need some additional notation. In particular we shall refer to (f-3) and (f-4) as follows:

(f-3)$|f(x,s)| \leq 2\gamma'|s| + f_o(x)$ $\forall$ $s \in \mathbf{R}$ and a.e. x$\in$ Ω where 0$<$ $\gamma' < \gamma$, $f_o \in L^2_\varrho(\Omega)$, and γ is defined in (2.10).

(f-4) $f(x,s) \geq -f_o(x)$ for $s \geq 0$ and $f(x,s) \leq f_o(x)$ for $s \leq 0$ and a.e. x$\in$ Ω where f_o is a nonnegative function in $L^2_\varrho(\Omega)$.

It is not difficult to see that if f meets (f-3) and (f-4) then

$$|f(x,s) - \gamma's| \leq \gamma'|s| + f_o(x) \;\; \forall s \in \mathbf{R} \;\; a.e. \; x \in \Omega. \qquad (2.15)$$

Hence (f-3) and (f-4) together imply (f-2).

In Theorem 5, the condition that we use in place of both (2.12) and (2.13) is (2.16) below where S_n is the subspace of $H^1_{p,q,\varrho}$ spanned by $\{\varphi_1, \ldots, \varphi_n\}$ with φ_j $(j = 1, \ldots, n)$ given in $(V_L - 1)$, $u_n \in S_n$, and $u_n = w_n + v_n$ with $< v_n, \varphi_j >= 0$ for $j = j_0, ..., j_0 + j_1 - 1$ and w_n a $\lambda_{j_o}-$eigenfunction of L. Also, we shall assume G$\in$ $[H^1_{p,q,\varrho}]'$, the dual of $H^1_{p,q,\varrho}$, that is, G: $H^1_{p,q,\varrho} \rightarrow \mathbf{R}$ linearly and $|G(u)| \leq K\|u\|_{p,q,\varrho} \forall u \in H^1_{p,q,\varrho}$ where K is a constant. The condition we then assume is the following :

Given $\{u_n\}^{\infty}_{n=j_0+j_1}$ with $u_n \in S_n$ and $u_n = v_n + w_n$. Suppose $\|u_n\|_\varrho \rightarrow \infty$ and $\|v_n\|_{p,q,\varrho} / \|u_n\|_{p,q,\varrho} \rightarrow 0$. Then

$$\limsup_{n\rightarrow\infty} \left[\left(1 - n^{-1}\right) < f(\cdot, u_n), w_n >_\varrho + G(w_n)\right] > 0. \qquad (2.16)$$

The result that we present concerning a generalization of the equation (2.9) using (2.16) is the following:

Theorem 5 *Assume all the conditions in the hypothesis of Theorem 4 excluding (2.12) and (2.13). Assume also that every $\lambda_{j_o}-$eigenfunction of L is* *a* $H^1_{p,q,\varrho}-\ \lambda_{j_o}-$ *eigenfunction of Q, that* $G \in [H^1_{p,q,\varrho}]'$*, that f meets (f-1),(f-3), and (f-4) , and that (2.16) holds. Then* $\exists\ u^* \in H^1_{p,q,\varrho}$ *which is a weak solution of*

$$Qu = [\lambda_{j_o} u + f(x,u)]\, \varrho - G \qquad (2.17)$$

We shall show in §3 that the above theorem is a best possible result in the following sense: if in (f-3) γ' is allowed to equal γ, then the above theorem is false. For semilinear results motivating Theorem 5 above, see [He] and [Rum].

By a weak a solution of (2.17), we shall mean

$\mathcal{Q}(u^*, v) = \lambda_{j_o} \langle u^*, v\rangle_\varrho + \langle f(\cdot, u^*), v\rangle_\varrho - G(v) \quad \forall v \in H^1_{p,q,\varrho}.$

Using some of the ideas in the proof of Theorem 5, we next assume the more restrictive hypothesis

$$(\mathrm{f}-5) \quad |f(x,s)| \le f_o(x)\, \forall s \in \mathbf{R}$$

where $f_o \in \mathrm{L}^2_\varrho(\Omega)$ and obtain using (f-5) , a theorem (Theorem 6 below) and a corollary along the lines of the two well-known resonance results of Landesman and Lazer [LaL] and Williams [Wi]. The corollary to this theorem will give a result which is both necessary and sufficient.

Because we make the more restrictive hypothesis (f-5), it turns out that in Theorem 6 we can weaken the condition in the hypothesis of Theorem 5 concerning the $H^1_{p,q,\varrho}-$ eigenvalue of Q. We do this by introducing the notion of an L-pseudo-eigenvalue of Q, and in §3 we will give examples of differential operators L and Q such that every eigenvalue of L is an L-pseudo-eigenvalue of Q. Hence for some L's and Q's, Theorem 6 will hold at every λ_{j_o}−eigenvalue of L.

We say λ_{j_o} is an L-pseudo-eigenvalue of Q provided the following obtains: (*i*) λ_{j_o} is an eigenvalue of L of multiplicity j_1; (*ii*)S_n and φ_j are as before in (2.16); (*iii*) $\mathrm{P}^o_{j_o}$u is the orthogonal projection of u onto the subspace of $\mathrm{H}^1_{p,q,\varrho}$ spanned by $\varphi_{j_o}, ..., \varphi_{j_o+j_1-1}$; (*iv*) if $\{u_n\}_{n=1}^\infty$ is a sequence with $u_n \in \mathrm{S}_n$, $\|u_n\|_{p,q,\varrho} \to \infty$, and $\left\|u_n - \mathrm{P}^o_{j_o} u_n\right\|_{p,q,\varrho} / \|u_n\|_{p,q,\varrho} \to 0$,

$$\text{then } \lim_{n\to\infty} \left[\, \mathcal{Q}\left(u_n, P^o_{j_o} u_n\right) - \mathcal{L}\left(u_n, P^o_{j_o} u_n\right)\,\right] / \|u_n\|_{p,q,\varrho} = 0 \qquad (2.18)$$

It is clear from the above that if every λ_{j_o}−eigenfunction of L is a $H^1_{p,q,\varrho}-$ λ_{j_o}−eigenfunction of Q, then λ_{j_o} is an L-pseudo-eigenvalue of Q.

For Theorem 6, 8, 11, and 12 we shall also need the following notation:

$$f_\pm(x) = \lim \inf_{s\to\pm\infty} f(x,s) \text{ and } f^\pm(x) = \lim \sup_{s\to\pm\infty} f(x,s) \qquad (2.19)$$

Theorem 6 *Let $\Omega \subset \mathbf{R}^N$, $N \geq 1$, be an open set, ϱ and $p_i(i = 1,\ldots, N)$ be positive functions in $C^0(\Omega)$ satisfying (1.1), $q \in C^0(\Omega)$ be a nonnegative function, and $\Gamma \subset \partial\Omega$ be a closed set. Let L and Q be given by (1.6) and (1.8) respectively and assume (Q-1) -(Q-5), that Q is #-related to L, and that (Ω, Γ) is a V_L- region. Suppose that f satisfies (f-1) and (f-5), that G $\in \left[H^1_{p,q,\varrho}\right]'$, and that λ_{j_o} is an eigenvalue of L of multiplicity j_1 and is also an L- pseudo-eigenvalue of Q. Suppose, furthermore, that*

$$G(w) < \int_{\Omega\cap[w>0]} f_+ w\varrho + \int_{\Omega\cap[w<0]} f^- w\varrho \tag{2.20}$$

for every nontrivial $\lambda_{j_o}-$ eigenfunction w of L. Then $\exists u^* \in H^1_{p,q,\varrho}$ which is a weak solution of (2.17).

Corollary 7 *Assume all the conditions in the hypothesis of Theorem 6 and in addition that every $\lambda_{j_o}-$ eigenfunction of L is a $H^1_{p,q,\varrho}-$ $\lambda_{j_o}-$ eigenfunction of Q. Assume also that*

$$f^-(x) < f(x,s) < f_+(x) \ \forall s \in \mathbf{R} \text{ and a.e. } x \in \Omega. \tag{2.21}$$

Then a necessary and sufficient condition that $\exists$ $u^* \in H^1_{p,q,\varrho}$ which is a weak solution of (2.16) is that (2.20) holds.

In Theorem 6, we obtained a Landesman-Lazer type result under the assumption (f-5) that $f(x,s)$ was uniformly bounded by a function in L^2_ϱ. If we introduce the notion of #-#-relationship, we can obtain a result of this nature at higher eigenvalues which allows $f(x,s)$ to be unbounded on one side (actually to satisfy a one-sided sublinear growth condition). With this in mind, we say Q is #-#-related to L if

(i) $\exists$ a constant K such that (2.22)

$$|\mathcal{Q}(u,v) - \mathcal{L}(u,v)| \leq K \quad \forall u,v \in H^1_{p,q,\varrho} \text{ with } \|v\|_{p,q,\varrho} \leq 1,$$

(ii) if $q(x)$is not identically zero, then $B_0(x) \geq a_0(x)$ in Ω.

Another way to view (i) in this #-#-relationship is to say

$$\mathcal{Q} - \mathcal{L} = O(1) \text{ uniformly for } \|v\|_{p,q,\varrho} \leq 1 \text{ as } \|u\|_{p,q,\varrho} \to \infty.$$

Also (ii) states the zeroth order term of Q dominates that of L.

It is clear that if Q is #-#-related to L, then it is #-related to L.

In Theorem 8, we shall be dealing with resonance from below; so (f-6) will be a different one-sided condition then (f-4).

(f-6) $f(x,s) \leq f_o(x)$ for s≥ 0, $f(x,s) \geq -f_o(x)$ for s ≤ 0 for a.e. $x \in \Omega$ where $f_o(x)$ is a nonnegative function in $L^2_\varrho(\Omega)$.

We shall also assume that $f(x,s)$ satisfies a sublinear growth condition, namely

(f-7) $\forall \varepsilon > 0, \exists b_\varepsilon \in L^2_\varrho(\Omega)$ $s.t.$ $|f(x,s)| \leq \varepsilon |s| + b_\varepsilon(x)$ $\forall s \in \mathbf{R}$ and a.e. x $\in \Omega$.

Motivated by [Sh 3], we present the following interesting theorem for f in (2.17) meeting the one-sided growth condition given by (f-6) and (f-7).

Theorem 8 *Let Ω, p_i, ϱ, q, G, L, Q, and Γ be as in the hypothesis of Theorem 6. Suppose in addition that Q is #-#-related to L, that f satisfies (f-1), (f-6), and (f-7) and that λ_{j_o} is an eigenvalue of L of multiplicity j_1 and also an L-pseudo-eigenvalue of Q. Suppose, furthermore, that*

$$G(w) > \int_{\Omega\cap[w>0]} f^+ w\varrho + \int_{\Omega\cap[w<0]} f\_w\varrho \tag{2.23}$$

for every nontrivial $\lambda_{j_o}-$ eigenfunction w of L. Then $\exists u^* \in H^1_{p,q,\varrho}$ which is a weak solution of (2.17).

Next, we are interested in improving significantly on Theorems 1 and 2 of $[Sh1]$. We are capable of doing this for some of the regions that were considered in that paper (e.g., the Bessel function example treated on the bottom of page 1414) and also for a large number of regions to be introduced in §3 of this monograph. We are mainly interested in obtaining superlinear improvements of both of the aforementioned theorems. What this is tantamount to is to replace the linear growth condition (f-3) in $[Sh1, p1413]$ by the superlinear growth condition

$$(f-8)\ \exists\theta \text{ with } 2 < \theta < 2N/(N-1) \text{ such that}$$

$$|f(x,s)| \leq h_o(x) + K|s|^{\theta-1} \quad \forall s \in \mathbf{R} \text{ and a.e. } x \in \Omega$$

where $h_o \in L^{\theta^*}_\varrho$, K is a nonnegative constant, and $\theta^* = \left(1-\theta^{-1}\right)^{-1}$.

An examination of the proofs given in $[Sh1]$ shows that what is needed to accomplish this is to obtain a new weighted compact Sobolev-type imbedding theorem. We succeed in doing this for a certain type of V_L-region which we shall label a Simple V_L-region and which we now define.

We assume Lu is of the form

$$Lu = -\sum_{i=1}^{N} D_i(p_i D_i u) + qu \tag{2.24}$$

where p=(p_1,...,p_n), q, and ϱ are as in the first paragraph of § 1. We say that (Ω,Γ) is a Simple V_L -region if (Ω,Γ) is a V_L-region and the following four conditions prevail:

(2.25)

(i) $\Omega = \Omega_1 \times ... \times \Omega_N$ where $\Omega_i \subset$ **R** is an open set for $i = 1, ..., N$;

(ii) associated with each Ω_i there are positive functions p_i^* and ϱ_i^* in $C^0(\Omega_i)$ satisfying $\int_{\Omega_i} [p_i^*(s) + \varrho_i^*(s)]\, ds < \infty$ for $i = 1, ..., N$;

(iii) $\varrho(x) = \varrho_1^*(x_1) ... \varrho_N^*(x_N)$ and

$$p_i(x) = \varrho_1^*(x_1) ... \varrho_{i-1}^*(x_{i-1})\, p_i^*(x_i)\, \varrho_{i+1}^*(x_{i+1}) ... \varrho_N^*(x_N)$$

for i= 1,..., N;

(iv) for each Ω_i $(i = 1, \dots, N)$, $\exists h_i \in C^0(\Omega_i) \cap L^\theta_{\varrho_i^*}(\Omega_i)$ for $2<\theta < \infty$ with the property that

$$\forall u \in C^1(\Omega_i), |u(s)| \leq h_i(s) \|u\|_{p_i^*, \varrho_i^*} \quad \forall s \in \Omega_i.$$

In (iv) above h_i is understood to be in every $L^\theta_{\varrho_i^*}(\Omega_i)$ for $2<\theta$, and also, to be quite explicit,

$$\|u\|^2_{p_i^*, \varrho_i^*} = \int_{\Omega_i} \left[p_i^*(s) |du(s)/ds|^2 + \varrho_i^* u^2(s)\right] ds. \tag{2.26}$$

It is easy to give interesting Simple V_L−regions. In particular the Bessel function region mentioned earlier in $[Sh1, p.1414]$ qualifies. Also the following joint Bessel - Legendre domain qualifies as a Simple V_L-region.

Take $\Omega = (0,1) \times (-1,1)$ with $\Gamma = \{ (1,s) : -1 \leq s \leq 1 \}$ and $p_1^*(x_1) = x_1, p_2^*(x_2) = (1 - x_2^2), \varrho_1^* = x_1, \varrho_2^* = 1, p_1(x) = x_1, \varrho(x) = x_1, p_2 = x_1(1 - x_2^2)$,and q$(x) = 0$.. It is clear that (2.25) $(i), (ii), (iii)$ hold for this example. All that remains to show is that (2.25)(iv) does indeed hold. We shall do this in §3 where we will also illustrate a number of further examples of Simple V_L−regions.

The new weighted compact Sobolev-type imbedding theorem that we shall establish is the following:

Theorem 9 . *Let Lu be given by (2.24) and suppose that (Ω,Γ) is a Simple V_L−region. Then for $N \geq 2$,$H^1_{p,q,\varrho}(\Omega,\Gamma)$ is compactly imbedded in $L^\theta_\varrho(\Omega)$ for every θ satisfying $2<\theta < 2N/(N-1)$. For N=1, $H^1_{p,q,\varrho}(\Omega,\Gamma)$ is compactly imbedded in $L^\theta_\varrho(\Omega)$ for every θ satisfying $2< \theta < \infty$.*

In the literature, there exists an excellent book on weighted Sobolev spaces by Kufner, [Ku]. Our weighted compact imbedding theorem, however, does not appear to be contained in this volume.

Using Theorem 9, we shall establish three superlinear results at the first eigenvalue, one of nonresonance type and two of resonance type. We deal, in particular with the following equation:

$$Qu = [\lambda_1 u + f(x,u)]\,\varrho - G \tag{2.27}$$

The first superlinear result we establish is

Theorem 10 *Let (Ω, Γ) be a Simple V_L-region where L is given by (2.24) and suppose Q, given by (1.8), satisfies (Q-1)-(Q-5) and is #-related to L. Suppose, also, that $G\in \left[H^1_{p,q,\varrho}\right]'$ and that $f(x,s)$ satisfies (f-1) and (f-8). Suppose, furthermore, $\exists\ \tilde{h}_o \in L^{\theta^*}_{\varrho}(\Omega)$ where $\tilde{h}_o$ is a nonnegative function and $\varepsilon^*_o > 0$ such that*

$$sf(x,s) \le -\varepsilon^*_o |s|^2 + \tilde{h}_o(x)\,|s| \quad \forall s \in \mathbf{R}^1 \textit{and a.e.} x \in \Omega \tag{2.28}$$

Then $\exists u^* \in H^1_{p,q,\varrho}$ which is a distribution solution of (2.27).

For the purposes of this monograph $u \in H^1_{p,q,\varrho}$ is a distribution solution of (2.27) will mean

$$\mathcal{Q}(u^*,v) = \lambda_1 < u^*, v >_\varrho + \int_\Omega f(x,u^*)\,v\varrho - G(v) \tag{2.29}$$

$\forall v \in H^1_{p,q,\varrho}$. The integral in (2.29) is indeed well-defined. To see this observe that f$(x,u^*)\,v \in L^1_\varrho(\Omega)$ follows from $(f-8)$, the continuous imbedding of $H^1_{p,q,\varrho}$ in $\mathrm{L}^\theta_\varrho(\Omega)$ given by Lemma 7 in §4, and Holder's inequality..

Observing that for $N=2$, $f(x,s) = -g(x)\,s\,|s|^{5/3} - \varepsilon^*_o s$, where g$(x) \in C^0(\Omega) \cap L^\infty(\Omega)$ is a positive function, meets both (f-8) and (2.28), we see that Theorem 10 is indeed a nonresonance superlinear result and does partially generalize [Sh1, Th1].

We also establish two resonance-type superlinear results. In order to accomplish this we replace (2.28) with

$$\exists \tilde{h}_o \in L^2_\varrho(\Omega) \text{ such that } sf(x,s) \le \tilde{h}_o(x)\,|s| \quad \forall s \in \mathbf{R} \text{ and a.e. } \mathrm{x} \in \Omega \tag{2.30}$$

where $\tilde{h}_o$ is a nonnegative function. Using (2.19), our first superlinear resonance-type result is the following.:

Theorem 11 *Assume all the conditions in the hypothesis of Theorem 10 except replace (2.28) with (2.30). Assume, furthermore,that Q is #-#-related to L and that* λ_1 *is an L-pseudo-eigenvalue of Q. Assume, also,*

$$\int_{\Omega} f^{+}\phi_1\varrho < G(\phi_1) < \int_{\Omega} f_{-}\phi_1\varrho \tag{2.31}$$

where ϕ_1 *is the positive first eigenfunction of L. Then* $\exists u^* \in H^1_{p,q,\varrho}$ *which is a distribution solution of (2.27).*

For a superlinear resonance theorem which is possibly a better extension of [Sh 1, Th.2], we need the following which replaces the L-pseudo-eigenvalue concept used in Theorem 11 above,

$$lim\ inf_{\|u\|_{p,q,\varrho}\to\infty}\ [\mathcal{Q}(u,u) - \mathcal{L}(u,u)]/\ \|u\|_{p,q,\varrho} \geq 0 \tag{2.32}$$

Our theorem then is

Theorem 12 *Assume all the conditions in the hypothesis of Theorem 10 except replace (2.28) with (2.30). Assume also that (2.31) and (2.32) hold. Then* $\exists u^* \in H^1_{p,q,\varrho}$ *which is a distribution solution of (2.27).*

For Simple V_L-regions and those Q's which are #-related to L, Theorem 12 is indeed a three-way extension of [Sh 1, Th.2]. For (i) it has the superlinear extension (f-8); (ii) it only assumes one-half of the *-relationship, namely (2.32), and does not assume the other half given in [Sh 1, (4.6)]; (iii) it can have a q $\neq$ 0 and different p_i's, none of which is allowed in [Sh 1, Th.2].

It is easy to give Simple V_L-regions and Q's for which Theorems 10,11, and 12 apply and we shall do so in the next section. In particular, every L which corresponds to a Simple V_L-region, itself qualifies for Theorems 10,11, and 12, and all three theorems are new even in the semi-linear case. The concluding example in the next section deals with a Q specifically covered by Theorem 12 and evidently not by Theorem 11.

To give some insight into the motivation for the theorems stated in this section, we observe the main idea is to study say an equation of the form

$$Qu = [\lambda u + g(x,u) + h(x)]\varrho \tag{2.33}$$

where $\lambda \in \mathbf{R}$, $g(x,s) \in L^{\infty}(\Omega \times \mathbf{R}) \cap C^0(\Omega \times \mathbf{R})$, h$\in L^2_{\varrho}(\Omega) \cap C^0(\Omega)$, and Q is our familiar singular quasilinear operator, and seek a weak solution u* of (2.33) in the Hilbert space $H^1_{p,q,\varrho}$. In [Sh1], for a special case of this Hilbert space, $H^1_{p,\varrho}$, a good result is obtained with respect to the equation in (2.33) for all $\lambda < \lambda_1^*$ where

$$\lambda_1^* = \liminf_{\|u\|_\varrho \to \infty} \mathcal{Q}(u,u)/ \|u\|_\varrho^2, \quad u \in H^1_{p,\varrho}. \tag{2.34}$$

Also, in case Q is asymptotically related to L in a special manner, a resonance result is obtained in [Sh1] for $\lambda = \lambda_1^*$. Therefore, results were known for the equation in (2.33) provided $\lambda \leq \lambda_1^*$. The question then becomes: *How does one proceed when* $\lambda > \lambda_1^*$? A number of theorems in this section are devoted specifically to the answering of this question.

In particular, if Q is #-related to L, we see with λ_1^* defined as in (2.34), that $\lambda_1^* = \lambda_1$, the first eigenvalue of L. Furthermore, if $\lambda_{j_o} < \lambda < \lambda_{j_o+j_1}$, then according to Remark 2 (which appears after the statement of Theorem 4) a weak solution u*$\in H^1_{p,q,\varrho}$ to the equation in (2.33) always exists. Likewise Theorems 4, 5, and 6 are devoted to finding a solution u*$\in H^1_{p,q,\varrho}$ to the equation in (2.33) when $\lambda = \lambda_{j_o}$, a higher eigenvalue of L. So the equation in (2.33) is covered for all values of $\lambda > \lambda_1^*$, and hence effectively for *all values of* $\lambda \in \mathbf{R}$.

Also, another motivation for the theorems in this section is to extend the type of singularity studied in [Sh1] so that it includes special functions of a more general type such as higher order Bessel functions [BD, p. 662] and the associated Legendre functions [CH, p. 327]. This is accomplished by introducing the weight q and by allowing the weights p_i to be different and working in the Hilbert space $H^1_{p,q,\varrho}$.

Finally , with G$\in [H^1_{p,\varrho}]'$ for the equation

$$Qu = [\lambda u + f(x,u)]\varrho - G \tag{2.35}$$

with $\lambda \leq \lambda_1$, there were no superlinear results in [Sh1]. On the other hand, results such as those in [dFG] for the semilinear situation when Q is replaced by L and the weights p_i, q,and ϱ are not present indicated that there might be some for the equation in (2.35) in the superlinear direction. We went looking for them and obtained the new weighted compact imbedding Theorem 9 and the superlinear Theorems 10, 11, and 12.

1.3 Examples

It is the purpose of this section to give various examples of the conditions that arise in the hypotheses of the ten theorems and the two corollaries that were stated in §2 of this chapter. In particular, using the theory of special functions, we first show a number of different V_L−regions, (Ω, Γ) (see under (1.7) for the definition of this concept). We are especially interested in those L's for which the weight $q \neq 0$ or for which the components of $p=(p_1 ..., p_N)$ are different, for these examples do not arise in the previous paper on this subject,[$Sh1$]. Next, we exhibit a number of quasilinear operators Q which are #-related and #-#-related to L (see under (1.10) and (2.22), respectively for these definitions). Also, we exhibit a number of situations where at a specific j_o every $\lambda_{j_o}-$ eigenfunction of L is a $H^1_{p,q,\varrho}-$ $\lambda_{j_o}-$eigenfunction of Q (see (2.2) for the definition). We also illustrate several situations where every eigenvalue of L is an L-pseudo-eigenvalue of Q (see (2.18)). In addition we exhibit a number of Simple V_L−regions (Ω, Γ) (see (2.25)) which are needed for Theorem 10,11, and 12. We also discuss several situations where the property (2.16) in Theorem 5 arises. For the quasilinear elliptic operator M, we exhibit several cases where an infinite number of near-eigenvalues arise. Throughout, special functions play a fundamental role, and we even use the associated Legendre functions, Jacobi polynomials, and higher order Bessel functions to illustrate the various new concepts that are presented in this chapter.

At the outset of the examples to be considered, we shall assume L is of the following more restrictive form

$$Lu = -\sum_{i=1}^{N} D_i \left[p_i D_i u\right] + qu \tag{3.1}$$

Later on, we shall deal with an L of the more general form (1.6).

Our first two examples will come from [$Sh1$]. With this in mind, take $\Omega = \mathbf{R}^2, \Gamma =$ empty set, $q = 0, p_1(x) = p_2(x) = \varrho(x) = e^{-\left(x_1^2 + x_2^2\right)}.$ Then

$$Lu = \sum_{i=1}^{2} D_i \left[\, e^{-\left(x_1^2+x_2^2\right)} D_i u \,\right],$$

and, as is well-known,

$$\Phi_{mn}(x_1, x_2) = H_m(x_1) H_n(x_2) / (2^{-(m+n)} m! n! \pi)^{\frac{1}{2}}, \tag{3.2}$$

where $H_n(t) = (-1)^n e^{t^2} d^n e^{-t^2} / dt^n$ is the familiar Hermite polynomial, gives rise to a CONS over Ω with respect to the weight $\varrho(x)$ for m,n = 0,1... . Since $L\ \Phi_{mn} = 2(m+n)\, \varrho \Phi_{mn}$, using the fact that C^1 functions vanishing outside of a compact are dense in $C^1_{p,q,\varrho}$, it is clear that $(V_L - 1)$ and $(V_L - 2)$ are satisfied with $\phi_1 = \Phi_{00}$ and $\lambda_1 = 0$. Hence (Ω, Γ) is a V_L−region.

For our second example, we work in $\mathbf{R}^3$, take $\check{\Omega} \subset \mathbf{R}^2$ to be a bounded connected open set, and $\Omega = \check{\Omega} \times \mathbf{R}^1$. So Ω is an infinite cylinder whose axis is parallel to the x_3-axis. Also, we take $\Gamma = \partial\check{\Omega} \times \mathbf{R}$, and L given by (3.1) with $p_1(x) = p_2(x) = p_3(x) = \varrho(x) = e^{-x_3^2}$, and $q(x) = 0$. Then letting $\{\psi_m(x_1, x_2)\}_{m=1}^{\infty}$ be the sequence of eigenfunctions for the Laplace operator in $W_0^{1,2}(\check{\Omega})$,with $\int_{\check{\Omega}} \psi_m^2 = 1$, and η_m the corresponding eigenvalue, we set

$$\Phi_{mn}(x) = \psi_m(x_1, x_2) H_n(x_3) / (2^{-n} n! \pi^{\frac{1}{2}})^{\frac{1}{2}} \tag{3.3}$$

for m=1,2,... and $n = 0, 1, 2, ...$ and observe that $\{\Phi_{mn}\}_{m=1,n=0}^{\infty,\infty}$ is a CONS over Ω with respect to the weight ϱ. Since L is of the form (3.1), we see that

$$L\Phi_{mn} = -e^{-x_3^2}[(D_1^2 + D_2^2)\Phi_{mn}] - D_3\left[e^{-x_3^2} D_3 \Phi_{mn}\right] = [\eta_m + 2n]\, \varrho \Phi_{mn}. \tag{3.4}$$

Consequently, with $\varphi_1 = \Phi_{10}$, we obtain from (3.4) that (V_L -2) is satisfied (see [GT, p. 214]). From (3.1), we see that ($V_L - 1$) is satisfied. Hence, (Ω, Γ) is a V_L−region.

Similarly, the third and fourth examples given in [*Sh*1, *p*. 1414] give rise to V_L−regions. In these last two examples, as well as those that arose in (3.2) and (3.3), we had all the p_i's equal and q=0. We now go on to consider situations where this is no longer the case. Hence we will be dealing with situations no longer covered by the assumption $O_{p,\varrho}(\Omega)$ in [Sh1].

For our first new example, we take $N = 2$, $\Omega = (-1, 1) \times (0, \pi)$, $\Gamma =$ the empty set. Also, we take $\varrho(x) = (1 - x_1)^{\alpha} (1 + x_1)^{\beta}$, $p_1(x) = (1 - x_1)^{\alpha+1} (1 + x_1)^{\beta+1}$, $p_2(x) = (1 - x_1)^{\alpha} (1 + x_1)^{\beta}$ where $\alpha, \beta > -1$, and $q = 0$. With L given by (3.1), we then have that

$$Lu = -D_1 (1 - x_1)^{\alpha+1} (1 + x_1)^{\beta+1} D_1 u - (1 - x_1)^{\alpha} (1 + x_1)^{\beta} D_2^2 u. \tag{3.5}$$

With $P_m^{\alpha,\beta}(t)$ representing the m-th Jacobi polynomial (see [*Ho*, *p*.40] or [*Ra*, *p*.254]), it is well-known that the sequence $\left\{P_m^{\alpha,\beta}(t)\right\}_{m=0}^{\infty}$ properly normalized forms a CONS on the interval (-1,1) with respect to the weight $(1 - t)^{\alpha} (1 + t)^{\beta}$. As a consequence,

$$\Phi_{mn}(x_1, x_2) = P_m^{\alpha,\beta}(x_1) \cos n x_2 \quad m, n = 0, 1, 2, ... \tag{3.6}$$

properly normalized forms a CONS over Ω with respect to the weight $\varrho(x)$. Also, from [*Ra*, *p*.258] and (3.6) above, we see that

$$L\Phi_{mn} = \left[m(1 + \alpha + \beta + m) + n^2\right] \varrho \Phi_{mn}. \tag{3.7}$$

Hence, with $\varphi_1 = \Phi_{00}$, we have that both $(\mathrm{V}_L - 1)$ and $(\mathrm{V}_L - 2)$ hold, and consequently (Ω, Γ) is a V_L-region. If we take $\Omega_1 = (-1, 1), \Omega_2 = (0, \pi)$, $\Gamma_1 = \Gamma_2 =$ the empty set, $\mathrm{p}_1^*(x_1) = (1 - x_1)^{\alpha+1}(1 + x_1)^{\beta+1}$, $\varrho_1^*(x_1) = (1 - x_1)^{\alpha}(1 + x_1)^{\beta}$, $p_2^*(x_2) = 1$, $\varrho_2^*(x_2) = 1$, $\mathrm{q}_1^* = \mathrm{q}_2^* = 0$, it is easy to see that (Ω, Γ) satisfies (2.25) $(i), (ii)$, and (iii). We shall also show that for -1< $\alpha, \beta \le 0$, (2.25) (iv) holds. Hence, the above region is a Simple V_L-region and qualifies for Theorems 10, 11, and 12.

For our second new example, we present one for which $q \neq 0$. We take $\mathrm{N} = 1$, $\Omega = (-1, 1)$, $\Gamma =$ the empty set, $\mathrm{p}(s) = (1 - s^2)$, $\varrho(s) = 1$, $\mathrm{q}(s) = (1 - s^2)^{-1}$. Then L given by (3.1) is of the form

$$Lu = -D_1(1 - s^2) D_1 u + (1 - s^2)^{-1} u \tag{3.8}$$

Now $\{ P_{n,1}(s)/a_n \}_{n=1}^{\infty}$, where $\mathrm{P}_{n,1}(s)$ is the first order associated Legendre function of degree n, with $a_n^2 = 2n(n+1)/(2n+1)$, gives rise to a CONS with respect to the weight $\varrho(s)$. Since (see [CH, p. 327, p. 512])

$$LP_{n,1}(s) = n(n+1)P_{n,1}(s) \quad n = 1, 2, ..., \tag{3.9}$$

it follows that both $(V_L - 1)$ and $(V_L - 2)$ hold. Hence (Ω, Γ) is a V_L-region. Since Ω is 1-dimensional, (2.25)$(i), (ii)$, and (iii) hold. We shall show later that (2.25)(iv) also holds. Hence (Ω, Γ) is, in addition, a Simple V_L-region.

For our third new example, we take $N = 2$, $\Omega = (-1, 1) \times (0, 1)$, $\Gamma = \{(t, 1) : -1 \le t \le 1\}$, $p_1(x) = (1 - x_1^2) x_2$, $p_2(x) = x_2$, $\varrho(x) = x_2$, $\mathrm{q}(x) = x_2(1 - x_1^2)^{-1} + x_2^{-1}$, and

$$Lu = -D_1(1 - x_1^2)x_2 D_1 u - D_2 x_2 D_2 u + \left[x_2(1 - x_1^2)^{-1} + x_2^{-1}\right] u. \tag{3.10}$$

Then

$$\Phi_{mn}(x_1, x_2) = P_{m,1}(x_1) J_1(k_n x_2) \quad m, n = 1, 2, ... \tag{3.11}$$

properly normalized forms a CONS on Ω with respect the weight $\varrho(x)$ where $\mathrm{P}_{n,1}$ is as in the previous example and J_1 is the familiar Bessel of the first kind of order 1 and k_n is the n-th positive root of $\mathrm{J}_1(t)$ (see $[Ho,\ p.\ 258]$). Now it follows from the well-known Bessel's equation $[BC,\ p.\ 336]$ and (3.8)–(3.11) above that

$$L\Phi_{mn} = \left[m(m+1) + k_n^2\right] \Phi_{mn} x_2. \tag{3.12}$$

We set $\varphi_1 = \Phi_{11}$ and as a consequence, we see from (3.12) and the previous observations that both $(V_L - 1)$ and $(V_L - 2)$ hold. Hence (Ω, Γ) is a V_L- region. Also, with $\Omega_1 = (-1,1), p_1^*(x_1) = (1 - x_1^2)$, $\varrho_1^*(x_1) = 1$, $\Omega_2 = (0,1)$, $p_2^*(x_2) = x_2$,and $\varrho_2^*(x_2) = x_2^*$, it follows that (Ω, Γ) satisfies (2.25) $(i), (ii)$, and (iii). We shall show later that both Ω_1 and Ω_2 also satisfy (2.25)(iv). Hence (Ω, Γ) qualifies as a Simple V_L-region for Theorems 10, 11, and 12.

For our fourth new example we present a one-dimensional region of infinite length that satisfies (2.25)(iv), and hence is a Simple V_L-region. In particular, we take $\Omega = (0, \infty)$, $\Gamma = \{0\}$, $p(s) = (s^2+1)^{-1}$, $\varrho(s) = e^{-s}$, and $\mathrm{q}(s) = 0$. Then with $\mathrm{D}_1 u = \mathrm{du/ds}$, it follows from (3.1) that

$$Lu = -D_1 \left(s^2+1\right)^{-1} D_1 u. \tag{3.13}$$

Writing $\mathrm{H}^1_{p,\varrho}$ for $\mathrm{H}^1_{p,q,\varrho}$ and setting $\|u\|^2_{p,\varrho} = \int_0^\infty [\, p\,|D_1 u|^2 + \varrho u^2\,]$ ds, we next see that $\mathrm{H}^1_{p,\varrho}$ is compactly imbedded in L^2_ϱ because of Ascoli-Arzela and the fact that

$$|u(x)| \leq \left[\int_0^x \left(s^2+1\right) ds\right]^{\frac{1}{2}} \left[\int_0^\infty p\,|D_1 u|^2\, ds\right]^{\frac{1}{2}} \tag{3.14}$$

for u$\in C^1_{p,\varrho}$. Also we see that the operator

$$T : L^2_\varrho \to H^1_{p,\varrho} \subset L^2_\varrho$$

defined by

$$\int_0^\infty pD_1 Tf D_1 v ds =< f, v >_\varrho \qquad \forall v \in H^1_{p,\varrho}$$

is a bounded, strictly positive, symmetric,completely continuous operator of L^2_ϱ into L^2_ϱ. Hence there exists a sequence of functions $\{\psi_m\}_{m=1}^\infty \in \mathrm{L}^2_\varrho \cap H^1_{p,\varrho}$ and $0 < \eta_1 < \eta_2 < \ldots < \eta_m \to \infty$ such that

$$\mathcal{L}(\psi_m, v) = \eta_m \langle \psi_m, v \rangle_\varrho \;\; \forall v \in H^1_{p,\varrho}.$$

Furthermore, $\{\psi_m\}_{m=1}^\infty$ is a CONS in $\mathrm{L}^2_\varrho\,(0,\infty)$. It turns out from the regularity theory involved here (see $[GF,\ p.\ 202]$) that ψ_m *is* in $\mathrm{C}^2(0,\infty)$ and from the uniqueness theory of second order ordinary differential equations that each eigenvalue η_m is simple and that $\psi_1(x) > 0$ for $0 < \mathrm{x} < \infty$. In addition, $\mathrm{L}\psi_m = \eta_m \psi_m \;\; \forall m$. Therefore $(\mathrm{V}_L - 1)$ and $(\mathrm{V}_L - 2)$ hold and (Ω, Γ) is a V_L-region.

To show that (Ω, Γ) is also a Simple V_L–region, let u(x) be in $C^1(\Omega)$ and suppose $\|u\|_{p,\varrho}$ is finite-valued. Then $u(x) - u(t) = \int_t^x D_1 u(s)ds$. hence

$$|u(x) - u(t)| \leq [\int_t^x (s^2+1)\, ds]^{\frac{1}{2}} \, \|u\|_{p,\varrho} .$$

Also,we see that

$$\left| u(x) - \int_0^\infty u(t)\, e^{-t} dt \right| \leq \int_0^\infty e^{-t} |u(x) - u(t)|\, dt.$$

As a consequence of these last two inequalities, we obtain after an easy computation that (2.25)(iv) holds with s replaced by x. Since (Ω, Γ) is a one dimensional V_L–region, (2.25)(i),(ii),and (iii) hold automatically, and (Ω, Γ) is indeed a Simple V_L–region.

For the next example, we take N=3, $\Omega = (-1,1) \times (0,1) \times (0,\infty)$,
$\Gamma = \{(r,i,t) : -1 \leq r \leq 1, 0 \leq t < \infty, i = 0,1\}$
$\cup \{(r,s,0) : -1 \leq r \leq 1, 0 \leq s \leq 1\}$,
$p_1 = (1 - x_1^2)\, e^{-x_3}, p_2 = e^{-x_3}$, $p_3 = (1 + x_3^2)^{-1}$,
$\varrho = e^{-x_3}$, and $q = (1 - x_1^2)^{-1}\, e^{-x_3}$,
and observe that

$$\Phi_{jkm}(x) = P_{j,1}(x_1)\ \sin k\pi x_2\ \psi_m(x_3) \quad j,k,m = 1,2,... \tag{3.15}$$

properly normalized, where ψ_m is the function in the previous example, forms a CONS in $L^2_\varrho(\Omega)$. (Recall, $P_{j,1}$ is the first order associated Legendre function of degree j.) Now with Lu given by (3.1), it follows from (3.9) and the remarks made in the previous example that

$$L\Phi_{jkm} = \left[j(j+1) + (k\pi)^2 + \eta_m \right] \varrho \Phi_{jkm} \tag{3.16}$$

It is clear from (3.15) that $(V_L - 1)$ holds. With $\varphi_1(x) = \Phi_{111}(x)$, it is also clear from (3.16) that $(V_L - 2)$ holds. Therefore (Ω, Γ) is a V_L–region. It is, furthermore, easy to see that (2.25)(i), (ii), and (iii) holds for (Ω, Γ). We shall show (2.25)(iv) holds for the region corresponding to (3.9). We already have shown that it holds for the region in the previous example. Consequently, putting these two facts together, we see that (2.25)(iv) holds for (Ω, Γ) in the current example. Hence, the (Ω, Γ) of our current example is also a Simple V_L -region.

Next, we show that the regions above which we said satisfy (2.25) (iv) actually do. An examination of various situations involved shows that there are essentially four one-dimensional cases to consider: Case A where Ω $=(-1,1)$, $p(s)=1-s^2$, and $\varrho(s)=1$; Case B where $\Omega=(0,1)$, $p(s)=s$, and $\varrho(s)=s$; Case C where $\Omega=(-1,1)$, $p(s)=(1-x)^{1+\alpha}(1+x)^{1+\beta}$, and $\varrho(s)=(1-s)^{\alpha}(1+s)^{\beta}$ $-1\leq s\leq 1$ and $-1<\alpha,\beta\leq 0$; Case D where $\Omega=(0,1)$, $p(s)=1$, and $\varrho(s)=1$.

We start out with Case A. We establish (2.25)(iv) by showing

$$\exists \text{positive constants } c_1 \text{ and } c_2 \text{ such that} \tag{3.17}$$

$$\left|u(s)-\frac{1}{2}\int_{-1}^{1}u(t)\,dt\right|\leq c_1\left[|\log(1+s)|^{\frac{1}{2}}+|\log(1-s)|^{\frac{1}{2}}+c_2\right]\|u\|_{p,\varrho}$$

for $s\in(-1,1)$ and $u\in C^1(\Omega)$ with $\|u\|_{p,\varrho}<\infty$ (see (2.26)).

To establish (3.17), we designate the left-hand side of the inequality in (3.17) by $I_u(s)$. Then it follows that

$$2I_u(s)\leq\left|\int_{-1}^{1}\left[\int_t^s(du/dr)\,dr\right]dt\right|\leq\|u\|_{p,\varrho}\int_{-1}^{1}\left|\int_t^s\left(1-r^2\right)^{-1}dr\right|^{\frac{1}{2}}dt \tag{3.18}$$

But $2\int_t^s\left(1-r^2\right)^{-1}dr=\log(1+r)/(1-r)\,|_s^t$ and it is easy to see that (3.17) follows from this computation and the inequalities in (3.18). From (2.26),$|\int_{-1}^{1}u(t)\,dt\leq 2^{\frac{1}{2}}\|u\|_{p,\varrho}$. Also,$|log(1+s)|^{\frac{1}{2}}, |log(1-s)|^{\frac{1}{2}}\in L_{\varrho}^{\theta}(-1,1)$ for every $\theta\in(2,\infty)$. Consequently, (2.25)(iv) follows from (3.17) and Case A is established.

For Case B, we establish (2.25)(iv) by showing

$$\exists \text{ exist positive constants } c_1 \text{ and } c_2 \text{ such that} \tag{3.19}$$

$$\left|u(s)-2\int_0^1 t\,u(t)\,dt\right|\leq c_1\left[|\log\ s|^{\frac{1}{2}}+c_2\right]\|u\|_{p,\varrho}$$

for $s\in(0,1)$ and $u\in C^1(\Omega)$ with $\|u\|_{p,\varrho}<\infty$.

To establish (3.19), we designate the left-hand side of the inequality in (3.19) by $I_u(s)$. Then it follows that

$$2^{-1}I_u(s)\leq\left|\int_0^1 t\left[\int_t^s(du/dr)\,dr\right]dt\right|\leq\int_0^1 t\left|\int_t^s r^{-1}dr\right|^{\frac{1}{2}}dt\,\|u\|_{p,\varrho} \tag{3.20}$$

But $\int_t^s r^{-1}dr$ = log s - log t and (3.19) follows from this fact and (3.20). From (2.26), $\left|\int_0^1 u(t)\, tdt\right| \leq 2^{-\frac{1}{2}} \|u\|_{p,\varrho}$. Also, $|\log\ s|^{\frac{1}{2}}$ is in $L_\varrho^\theta(0,1)$ for every $\theta \in (2,\infty)$. Consequently, (2.25)(iv) follows from (3.19), and Case B is established.

Using the inequality,

$$(1-s)^{-(1+\alpha)}(1+s)^{-(1+\beta)} \leq (1-s)^{-(1+\alpha)} + (1+s)^{-(1+\beta)}$$

for $-1 < s < 1$, $-1 < \alpha, \beta \leq 0$, we see that Case C follows in a manner very similar to Case A. Case D is easy. We leave the details of both these cases to the reader.

It follows from Case A, Case B, Case C and Case D that indeed all the regions that up to now we have said are Simple V_L-regions actually are.

Next, we give three examples where Lu is of the general form (1.6) but not necessarily of the more restricted form (3.1). We take the obvious example first, namely

$$Lu = -\sum_{i=1}^{N} D_i \sum_{j=1}^{N} a_{ij}(x) D_j u + a_0(x) u \tag{3.21}$$

where $\Omega \subset \mathbf{R}^N$ is bounded open connected set, $a_{ij}(x) \in C^3(\Omega) \cap L^\infty(\Omega)$, $a_0(x) \in C^1(\Omega) \cap L^\infty(\Omega)$, $p_i(x) = 1$, $i = (1,...,N)$, $q(x), \varrho(x) = 1$, and the conditions in (1.5) holds. Also, we take $\Gamma = \partial\Omega$. Then $H^1_{p,q,\varrho}$ becomes the familiar $W_0^{1,2}(\Omega)$, and (Ω,Γ) is a V_L–region (see [*GT*, *p.* 214] and [*Hel*, *p.* 199].).

For our next to final example, we work in $\mathbf{R}^3$, take $\check{\Omega} \subset \mathbf{R}^2$ to be a bounded open connected set, $\Omega = \check{\Omega} \times (-1,1)$ and $\Gamma = \partial\check{\Omega} \times [-1,1]$. Also, we take $p_1 = p_2 = \varrho = 1$, $q = 0$, $p_3(x) = (1-x_3^2)$ and $a_{ij} \in C^\infty(\check{\Omega}) \cap L^\infty(\check{\Omega})$ and meeting, in addition, the conditions in (1.5) with $\check{\Omega}$ replacing Ω. Operating at first in $\check{\Omega}$, we set

$$L^\flat w(x_1,x_2) = -\sum_{i=1}^{2} D_i \sum_{j=1}^{2} a_{ij}(x_1,x_2) D_j w(x_1,x_2)\ ,$$

and once again (as in the previous example with $N = 2$ and now $q = 0$) we obtain a CONS, $\{\psi_m(x_1,x_2)\}_{m=1}^\infty$, in $L^2(\check{\Omega})$ which is also in $W_0^{1,2}(\check{\Omega}) \cap C^\infty(\check{\Omega})$ (see [*GT*, *p.* 214].). We designate the corresponding sequence of eigenvalues by $\{\eta_m\}_{m=1}^\infty$. Next we define

$$Lu(x_1,x_2,x_3) = -L^\flat u(x_1,x_2,x_3) - D_3(1-x_3^2) D_3 u(x_1,x_2,x_3) \tag{3.22}$$

for u$\in C^1_{p,q,\varrho}(\Omega)$. Then it is clear that

$$\left\{\psi_m(x_1,x_2)\,P_n(x_3)\,(2n+1)^{\frac{1}{2}}/\sqrt{2}\right\}_{m=1,n=0}^{\infty,\infty}$$

is a CONS for $L\varrho^2(\Omega)$ where P_n is the n-th Legendre polynomial. Furthermore, we have

$$L\psi_m(x_1,x_2)\,P_n(x_3) = [\eta_m + n(n+1)]\,\psi_m P_n.$$

Hence, (Ω,Γ) is a V_L−region and our next to final example is complete.

For our final example, once again N=3, and we take $\check{\Omega}$ as in the last example. Also, we take $p_1 = p_2 = p_3 = \varrho = 1$, $q = 0$, and $a_{ij} \in C^\infty(\check{\Omega}) \cap L^\infty(\check{\Omega})$, and in addition meeting the conditions in (1.5). We define $L^\flat w$ as in the last example and have $\{\psi_m(x_1,x_2)\}_{m=1}^{\infty}$and $\{\eta_m\}_{m=1}^{\infty}$as before with $L^\flat\psi_m = \eta_m\psi_m$ and $0 < \eta_1 < \eta_2 \leq ... \leq \eta_m \to \infty$. Suppose η_{m_o} is the first eigenvalue of $L^\flat$ which is strictly greater than η_2. We then take $\alpha = \pi/(\eta_{m_o} - \eta_1)^{\frac{1}{2}}$ and $\Omega = \check{\Omega} \times I_\alpha$ where I_α is the open interval (0,α). Also, we take

$$\Gamma = \{\,(x_1,x_2,x_3) : (x_1,x_2) \in \partial\check{\Omega} \text{ and } x_3 \in \bar{I}_\alpha\,\}$$

We then observe that $\{\Phi_{mn}\}_{m=1,n=1}^{\infty,\infty}$ where

$$\Phi_{mn} = \psi_m(x_1,x_2)\;\;\beta_n\cos[\,(n-1)\,\pi\alpha^{-1}x_3\,]$$

gives rise to CONS in $L^2_\varrho(\Omega)$ for $\beta_1 = \sqrt{1/\alpha}$ and $\beta_n = \sqrt{2/\alpha}$ for n$\geq$ 2. Furthermore, we define

$$Lu = L^\flat u - D_3^2 u \;\;\forall u \in C^1_{p,\varrho}(\Omega,\Gamma), \tag{3.23}$$

and observe that

$$L\Phi_{mn} = \left[\eta_m + (n-1)^2\,\pi^2\alpha^{-2}\right]\Phi_{mn} \tag{3.24}$$

Consequently, (Ω,Γ) is a V_L-region and our final example of such regions is complete. Also, we note for future use that the first $(m_0 - 1)$ eigenvalues of L are $\eta_1, ..., \eta_{m_o-1}$ with corresponding eigenfunctions $\psi_1, ..., \psi_{m_o-1}$.

Next, we give a number examples of Q's which are #-related and #-#-related to L and for which the condition concerning eigenvalues in Theorems 1 and 5 and Corollaries 2 and 7 are valid, namely that every λ_{j_o}-eigenfunction of L is also a $H^1_{p,q,\varrho}$− λ_{j_o}-eigenfunction of Q. Also, some of these examples will qualify for Theorems 4, 6, 8, 10, 11 and 12. In these examples, we shall assume at the outset that L is of the form (3.1). Later on we give examples relating Q to the more general form (1.6).

We start out with the case governed by the example corresponding to (3.2). In this case with Q defined by (1.8) , we take $B_0 = 0$ and

$$A_i(x,t,\xi) = \left\{1 + \left[1 + |\xi|^2\right]^{-\frac{1}{2}}\right\} \xi_i \; p_i^{\frac{1}{2}} \; i = 1,2. \tag{3.25}$$

It is then clear from $[KS, p.96]$ that Q meets (Q-1)-(Q-5) and from (3.1) and (1.10) that

$$\mathcal{Q}(u.v) - \mathcal{L}\,(u,v) = \int_{\mathcal{R}^2} e^{-|x|^2} \left[\sum_{i=1}^{2} D_i u D_i v\right] \left[1 + |Du|^2\right]^{-\frac{1}{2}} \tag{3.26}$$

where $Du = (D_1 u,\ D_2 u)$. If $\|v\|_{p,q,\varrho} \le 1$, then it follows from Schwarz's inequality and (3.26) that

$$|\mathcal{Q}(u,v) - \mathcal{L}\,(u,v)| \le \pi^{\frac{1}{2}} \qquad \forall u,v \in H^1_{p,q,\varrho}. \tag{3.27}$$

It is clear from this last inequality that (1.11) (i) is satisfied. Since $q = 0$, (1.11) (ii) is also satisfied. Hence Q given by (3.25) is #-related to L. It is also clear from (3.27) and (2.22) that Q is #-#-related to L.

We also observe from the calculation involved with (3.2) that the first eigenvalue for the corresponding L is $\lambda_1 = 0$. Furthermore, it is clear that this eigenvalue is simple and every λ_1- eigenfunction is a constant multiple of $H_0(x_1)\, H_0(x_2)$, *i.e.*, a constant. Also, we see from (3.26) that $Q(v,c) = 0\ \forall v \in H^1_{p,q,\varrho}$ and for every constant c. Hence, it follows that every λ_1- eigenfunction of L is also a $H^1_{p,q,\varrho}-\ \lambda_1-$ eigenfunction of Q. This example therefore qualifies for Theorems 1, 5, 6, and 8 and Corollary 7 when $\lambda_{j_o} = \lambda_1$. It also qualifies at every eigenvalue of L, i.e., at $\lambda_j\ \forall\ j$, for Theorem 4 since (Ω, Γ) is a V_L-region and Q is #-related to L.

Next, we give an example of a Q and an L for which an infinite number of different λ_{j_o} 's of L have all their eigenfunctions as $H^1_{p,q,\varrho}-$ eigenfunctions of Q. For this, we use the example where the eigenfunctions are given by (3.3) and where the η_m in (3.4) are all irrational and η_m - $\eta_{m'}$ is also irrational if $\eta_m \ne \eta_{m'}$. (For example, we can take $\check{\Omega}$ to be the square (0,1) x (0,1).) Hence it follows that

$$n \text{ is a positive integer} \Rightarrow \eta_{m'} + 2n \ne \eta_m\ \forall \eta_m \ne \eta_{m'}. \tag{3.28}$$

We next assume that the real-valued function F(t) has the following properties:

$$(i)\ F(t) \in C^0([0,\infty)) \text{ is nondecreasing and positive for t>0;} \qquad (3.29)$$

$$(ii)\ \lim_{t\to\infty} t\,[1-F(t)] = 0,$$

e.g., $F(t) = t^2/\left(1+t^2\right),\ t/\sqrt{1+t^2},\ \left[(1+t)\,/\left(2+t^2\right)\right]^{3/2}$, etc.

The $A_i(x,s,\xi)$ of Q in (1.8) are then defined to be

$$A_1(x,s,\xi) = p_1^{\frac{1}{2}}\xi_1,\ A_2(x,s,\xi) = p_2^{\frac{1}{2}}\xi_2$$

$$A_3(x,s,\xi) = 2^{-1}\left[\,1+F(|\,\xi_3\,|)\,\right]\xi_3 p_3^{\frac{1}{2}}$$

where $p_1 = p_2 = p_3 = e^{-x_3^2} = \varrho$ and $q = 0$.

It is then clear with this definition that Q meets (Q-1)-(Q-5) and that

$$\mathcal{Q}(u,v) - \mathcal{L}(u,v) = -2^{-1}\int_\Omega e^{-x_3^2}\left[1-F\left(|D_3u|\right)\right]D_3uD_3v. \qquad (3.30)$$

Now it follows from (3.29) that there is a constant c_1 such that $|1-F(t)|\,t \le c_1\ \forall t \in (0,\infty)$. Consequently, if $\|v\|_{p,q,\varrho} \le 1$, we obtain that

$$|\mathcal{Q}(u,v) - \mathcal{L}(u,v)| \le c_1 2^{-1}\left[\int_\Omega e^{-x_3^2}\right]^{\frac{1}{2}} \quad \forall u \in H^1_{p,q,\varrho}$$

That Q is #-#-related to L then follows immediately from this last inequality and (2.22).

Next we see from (3.4) and (3.28) that the only η_m- eigenfunctions of L are finite linear combinations of functions of the form Φ_{m0}. We claim that every such function is an $H^1_{p,q,\varrho}-\ \eta_m-$ eigenfunction of Q. To see this, observe that $\Phi_{m0}(x) = \psi_m(x_1,x_2)\ H_0(x_3)$ where $H_0(x_3)$ is a constant. Therefore, $D_3\ \Phi_{m0} = 0$ and it follows from (3.30) that for $v \in H^1_{p,q,\varrho}$

$$\mathcal{Q}(v,\Phi_{m0}) = \mathcal{L}(v,\Phi_{m0}) = \eta_m < v,\Phi_{m0} >_\varrho .$$

So every $\lambda_{j_o}-$ eigenfunction of L is a $H^1_{p,q,\varrho}-\ \lambda_{j_o}$ - eigenfunction of Q when $\lambda_{j_o} = \eta_m$ and there are an infinite number of such λ_{j_o}'s. This example qualifies for Theorems 1,4,5,6.and 8 as well Corollary 7.

It is easy to see that a similar situation prevails for the example in $\mathbf{R}^2$ with the eigenfunctions given by (3.6). In particular, it follows from (3.7) that if $\alpha + \beta$ is irrational, then with Q defined by

$$Qu = Lu - \; 2^{-1} (1 - x_1)^{\alpha} (1 + x_1)^{\beta} D_2 [F(|\, D_2 u \,|) - 1] \, D_2 u$$

with L given by (3.5) and F by (3.29), every λ_{j_o} - eigenfunction of L is a $H^1_{p,q,\varrho} - \lambda_{j_o} -$ eigenfunction of Q when $\lambda_{j_o} = m(\,1 + \alpha + \beta + m)$, m $= 0, 1, ...$. This example qualifies for Theorem 1, 4, 5, 6, and 8 and Corollaries 2 and 7. It also qualifies for Theorems 10, 11, and 12 when -1 $< \alpha, \beta \leq 0$. It qualifies for Theorem 12 because using the same technique as in (3.34) below

$$lim_{\|u\|_{p,q,\varrho} \to \infty} [\mathcal{Q}(u,u) - \mathcal{L}(u,u)] \, / \, \|u\|_{p,q,\varrho} = 0$$

Hence, (2.32) holds.

We next give three examples of an L and a Q where every eigenvalue λ of L is an L - pseudo-eigenvalue of Q and Q is #-#-related to L. Hence these examples will qualify for Theorems 6 and 8 (as well as Theorem 4). The first two examples also qualify for Theorems 10 and 11.

For our first such example, we take Lu given by (3.10) and Qu defined by

$$Qu = Lu - 2^{-1} D_2 \{\, x_2 \, [F(|D_2 u|) - 1] \, D_2 u \,\} \tag{3.31}$$

where F meets the conditions in (3.29). As a consequence, it follows that

$$\mathcal{Q}(u,v) - \mathcal{L}(u,v) = 2^{-1} \int_{\Omega} x_2 \, [F(|D_2 u|) - 1] \, D_2 u D_2 v \tag{3.32}$$

Since there is a constant c such that $|[\, F(t) - 1\,] t| \leq c \;\forall t \in (0, \infty)$, it follows that for $\|v\|_{p,q,\varrho} \leq 1$,

$$|\mathcal{Q}(u,v) - \mathcal{L}(u,v)| \leq c \, 2^{-1} \left[\int_{-1}^{1} dx_1 \int_0^1 x_2 dx_2 \right]^{\frac{1}{2}} = c_3$$

Hence Q is #-#-related to L.

Next, let λ_{j_o} be an eigenvalue of L of multiplicity j_1 and suppose that $\varphi_{j_o}, \varphi_{j_{o+1}}, ..., \varphi_{j_o + j_1 - 1}$ are the orthonormal eigenfunctions that correspond to this eigenvalue. Also let $P^o_{j_o} u$ be the orthogonal projection of u onto the subspace of $H^1_{p,q,\varrho}$ spanned by $\varphi_{j_o}, ..., \varphi_{j_o + j_1 - 1}$. Furthermore, let $\{u_n\}_{n=1}^{\infty}$ be sequence of functions as described in (2.18)(iv), i.e., $\|u_n\|_{p,q,\varrho} \to \infty$

and $\left\|u_n - P^o_{j_o}u_n\right\|_{p,q,\varrho} / \|u_n\|_{p,q,\varrho} \rightarrow 0$. In order to show that λ_{j_o} is an L-pseudo-eigenvalue of Q, the following has to be established:

$$\lim_{n\rightarrow\infty} \left|\mathcal{Q}\left(u_n, P^o_{j_o}u_n\right) - \mathcal{L}\left(u_n, P^o_{j_o}u_n\right)\right| / \|u_n\|_{p,q,\varrho} = 0 \tag{3.33}$$

Choosing t_0 such that $|1 - F(t)|\,|t| \leq \varepsilon$ for $t \geq t_0$, it follows from (3.32) that

$$| \mathcal{Q}(u_n, u_n) - \mathcal{L}(u_n, u_n) | \leq 2^{-1}t_0^2 | \int_\Omega x_2 dx |$$

$$+2^{-1}\varepsilon \|u_n\|_{p,q,\varrho} \left[\int_\Omega x_2 dx_2\right]^{\frac{1}{2}}.$$

Hence it follows that

$$\lim_{n\rightarrow\infty} | \mathcal{Q}(u_n, u_n) - \mathcal{L}(u_n, u_n) | / \|u_n\|_{p,q,\varrho} = 0 \tag{3.34}$$

Next, we set $v_n = u_n - P^o_{j_o}u_n$ and observe from the inequality below (3.32) that $|\mathcal{Q}(u_n, v_n) - \mathcal{L}(u_n, v_n)| \leq c_3 \|v_n\|_{p,q,\varrho}$. Therefore from the conditions in the hypothesis of (2.18)(iv),

$$\lim_{n\rightarrow\infty} |\mathcal{Q}(u_n, v_n) - \mathcal{L}(u_n, v_n)| / \|u_n\|_{p,q,\varrho} = 0.$$

However,

$$\begin{aligned} \mathcal{Q}\left(u_n, P^o_{j_o}u_n\right) - \mathcal{L}\left(u_n, P^o_{j_o}u_n\right) &= \mathcal{Q}(u_n, u_n) - \mathcal{L}(u_n, u_n) \\ &\quad +\mathcal{L}(u_n, v_n) - \mathcal{Q}(u_n, v_n) \end{aligned}$$

Consequently, from (3.34) and this last limit, it follows that (3.33) holds. Therefore, λ_{j_o} is an L-pseudo-eigenvalue of Q, and our first example for Theorems 6 and 8, as well as Theorems 10 and 11, is complete.

For our second example, we take the weights for L defined by (3.1) as in the example corresponding to (3.15) and we define Q as follows:

$$Qu = Lu - 2^{-1}D_1 \{ p_1 D_1 [F(|D_1 u|) - 1] D_1 u \} \tag{3.35}$$

where F meets the conditions in (3.29). As a consequence it follows that

$$\mathcal{Q}(u, v) - \mathcal{L}(u, v) = 2^{-1} \int_\Omega (1 - x_1^2)\, \varrho^{-x_3} [F(|D_1 u|) - 1] D_1 u D_1 v \tag{3.36}$$

Since there is a constant c such that $\|[\,F(t)-1\,]\,t| \leq c$, it follows from (3.36) that for $\|v\|_{p,q,\varrho} \leq 1$,

$$|Q(u,v)\text{-}\mathcal{L}(u,v)| \leq c2^{-1}\left[\int_0^{\infty}\int_0^1\int_{-1}^1 \left(1-x_1^2\right)e^{-x_3}dx_1dx_2dx_3\right]^{\frac{1}{2}} = c_4$$

Hence, Q is #-#-related to L.

The rest of this example follows in the exact same manner as the previous example did, and we leave the details to the reader.

Next, we give an example which is #-#-related to an L of the more general form (1.6). In particular, we take L to be given by (3.23) and then define Q by

$$Qu = Lu - 2^{-1}D_3\left[F\left(|D_3u|\right)-1\right]D_3u \tag{3.37}$$

where F meets the conditions in(3.29). It then follows that

$$\mathcal{Q}(u,v) - \mathcal{L}(u,v) = 2^{-1}\int_{\Omega}\left[F\left(|D_3u|\right)-1\right]D_3uD_3v, \tag{3.38}$$

and exactly the same arguments that worked for the previous two examples show that Q is #-#-related to L and that every eigenvalue λ of L is also an L-pseudo-eigenvalue of Q. So this example has relevance for Theorems 6 and 8 (and also Theorem 4). Also this example has relevance for Theorems 1 and 5 because it shows that there is more than one eigenvalue λ_{j_o} such that every $\lambda_{j_o}-$ eigenfunction of L is also a $H^1_{p,q,\varrho}-\ \lambda_{j_o}-$ eigenfunction of Q. To be precise, let $m_0 > 2$ be the positive integer defined in the discussion regarding (3.23). Then the first (m_0-1) eigenvalues of L are $\eta_1, ..., \eta_{m_o-1}$ with corresponding eigenfunctions $\psi_1(x_1,x_2), ..., \psi_{m_0-1}(x_1,x_2)$. We claim that ψ_j is also an $H^1_{p,q,\varrho}-$ eigenfunction for Q, j=1,..., m_0-1.

To see this, we observe that $\mathcal{L}\ (u,\psi_j) = \eta_j <$ u, $\psi_j >_{\varrho}$, that $\Omega = \check{\Omega} \times (0,\alpha)$ and that

$$\int_0^{\alpha}\int_{\hat{\Omega}}\left[F\left(|D_3u|\right)-1\right]D_3uD_3\psi_j\left(x_1,x_2\right)dx_1dx_2dx_3 = 0.$$

Hence from (3.38), it follows that $\mathcal{Q}(u,\psi_j) = \eta_j <$ u, $\psi_j >_{\varrho}\ \forall$ u $\in H^1_{p.q,\varrho}$, and our assertion that ψ_j is an $H^1_{p,q,\varrho}-$ eigenfunction for Q, $j = 1, ..., m_0 - 1$ is indeed true.

We next exhibit two different examples of an L and M with an infinite number of near-eigenvalues (see (2.6)). For both examples we take N = 3 and $\sigma = (\sigma_0, \sigma_1, \sigma_2, \sigma_3)$ with σ_i meeting $(\sigma-1)-(\sigma-4)$ for i= 0, 1, 2 and

σ_3 meeting $(\sigma - 1) - (\sigma - 3)$. Also, we assume $\sigma_0 = \sigma_1 = \sigma_2 \neq \sigma_3$. (It is clear that in case $\sigma_0 = \sigma_1 = \sigma_2 = \sigma_3$, then every eigenvalue of L is a near-eigenvalue of M provided σ_1, meets $(\sigma - 4)$.)

For the first example, we take q $= 0$ and L given by (3.1) with $\varrho = p_1 = p_2 = p_3 = e^{-x_3^2}$ and $\Omega = \check{\Omega} \times \mathbf{R}^1$ with $\check{\Omega} = (0,1) \times (0,1)$,the unit square. Then

$$\mathcal{L}(u,v) = \int_\Omega \sum_{i=1}^{3} e^{-x_3^2} D_i u D_i v \tag{3.39}$$

with the eigenfunctions of L given by (3.3). We next set

$$M = -\sum_{i=1}^{3} D_i \left[p_i \sigma_i (u) D_i u \right] \quad \text{with } \sigma_1 = \sigma_2 \neq \sigma_3.$$

From (3.3) and (3.4) we see that

$$\mathcal{L}(u, \Phi_{m0}) = \eta_m \langle u, \Phi_{m0} \rangle_\varrho \quad \forall u \in H^1_{p,q,\varrho} \tag{3.40}$$

where the η_m satisfy (3.28) and therefore are eigenvalues of L whose only eigenfunctions are finite linear combinations of functions of the form Φ_{m0}. Since $D_3 \Phi_{m0} = 0$, it follows that

$$\mathcal{M}(u, \Phi_{m0}) = \int_\Omega \sum_{i=1}^{2} e^{-x_3^2} \sigma_i (u) D_i u D_i \Phi_{m0} .$$

Consequently, we see from (3.39) and (3.40) that

$$\mathcal{M}(u, \Phi_{m0}) = \sigma_1 (u) \mathcal{L}(u, \Phi_{m0}) = \sigma_1 (u) \eta_m \langle u, \Phi_{m0} \rangle_\varrho \ \forall u \in H^1_{p,q,\varrho}.$$

Hence, it follows from (2.7) that η_m is a near-eigenvalue of M. Since there are an infinite number of such η_m , namely every number of the form $\pi^2 \left(m_1^2 + m_2^2\right)$ where m_1 and m_2 are positive integers, our first example for near-eigenvalues is complete.

For the next example, q$\neq 0$. We take N $= 3$, $\Omega = (-1,1) \times (0,1) \times (-\infty, \infty)$, $p_1 = e^{-x_3^2}\left(1 - x_1^2\right)$, $p_2 = e^{-x_3^2}$, $p_3 = e^{-x_3^2}$, $q = e^{-x_3^2}\left(1 - x_1^2\right)^{-1}$, $\varrho = e^{-x_3^2}$, and L given by (2.1). Also we take

$$\Gamma = \{(t,i,s) : -1 \le t \le 1, -\infty < s < \infty, i = 0,1\}.$$

It then follows that

$$\Phi_{jkm}(x) = P_{j,1}(x_1) \ sin \ k\pi x_2 \ H_m (x_3) \quad j,k = 1,2...,m = 0,1,2, \tag{3.41}$$

properly normalized forms a CONS on Ω with respect to the weight ϱ and that the eigenvalues of L are of the form $j(j+1)+(k\pi)^2+2m$. Hence $2+(k\pi)^2$ is a simple eigenvalue with eigenfunction Φ_{1k0}. It is an easy matter to check using (3.41) that (V_L-1) and (V_L-2) hold. Hence (Ω,Γ) is a V_L - region

We take

$$\mathcal{M}(u,v)=\int_\Omega\left[\sum_{i=1}^{3}p_i\sigma_i(u)\,D_iuD_iv+\sigma_0(u)\left(1-x_1^2\right)^{-1}e^{-x_3^2}uv\right] \qquad (3.42)$$

where $\sigma_0=\sigma_1=\sigma_2\neq\sigma_3$. Since $H_0(x_3)$ is a constant, it follows from (3.41) and (3.42) that

$$\mathcal{M}(u,\Phi_{1k0})=\sigma_1(u)\,\mathcal{L}(u,\Phi_{1k0})=\sigma_1(u)\left[2+(k\pi)^2\right]\langle u,\Phi_{1k0}\rangle_\varrho$$

for $u\in H^1_{p,q,\varrho}$ where $k=1,2,\ldots$. Since σ_1 meets $(\sigma-4)$, it follows from this last computation that $2+(k\pi)^2$ is a near-eigenvalue for M for k = 1,2,..., and our second example for an infinite number of near-eigenvalues is complete.

We are next concerned with various examples that arise in connection with Theorems 4 and 5. First of all we will give an example of an L (with L = Q also) that meets all the conditions in the hypothesis of Theorem 4 except (2.13), but for which the conclusion of Theorem 4 with L=Q is false.

We choose the simplest possible example to illustrate this situation and take N = 1, $\Omega=(0,\pi)$, $\Gamma=\{0,\pi\}$, p = 1, q = 0, $\varrho=1$, and $a_{11}=1$. Therefore, in (2.9) we take Qu = Lu = - D_1D_1u. Also, we take $j_o=1$, $j_1=1$, and f(x,s) = sin x. Since $\lambda_1=1$ and $\lambda_2=4$, we have by (2.10) that $\gamma=3/2$. Consequently, f(x,s) meets (f-1) and (f-2) with $f_o(x)=|\sin x|$. Now, it follows from (2.11) that both $\mathcal{F}^\pm(x)=0$ and $\mathcal{F}_\pm(x)=0$. Therefore, we see that (2.12) holds but (2.13) does not. So all the conditions in the hypothesis of Theorem 4 hold except for (2.13). Now the equation in (2.9) becomes $-D_1D_1u=u+\sin x$, and if a solution $u^*\in H^1_{p,q,\varrho}$ existed for this equation, we would have by (2.14) that

$$\int_0^\pi D_1u^*D_1v=\int_0^\pi[u^*+\sin x]v \qquad \forall v\in H^1_{p,q,\varrho}.$$

If this were indeed true, then on taking v = sin x in this last equation we would have that $0=\pi/2$, a contradiction! Hence no solution to (2.9) exists in $H^1_{p,q,\varrho}$, and our example is complete.

Next, we give an example with L=Q where all the conditions in Theorem 4 hold except for (2.13), where the function $f(x,s)$ has linear growth in s, and where $f(x,s)$ meets (2.16) with G = 0 and (f-2), (f-3), and (f-4); so that a solution to the equation in (2.9) actually exists by Theorem 5. Hence, this is an example of a situation that Theorem 4 (as well as Theorem 6 and Theorem 8 also) does not cover, but Theorem 5 does cover.

Once again we choose a simple situation to illustrate this case. We take $N = 1$, $\Omega = (0,\pi), \Gamma =$ the empty set, $p = \varrho = 1$, $q = 0$, $j_o = 1, j_1 = 1$, and $Qu = Lu = -D_1 D_1 u$. Then, as is well-known, $\varphi_1(x) = \pi^{-\frac{1}{2}}$, $\varphi_n(x) = 2^{\frac{1}{2}}\pi^{-\frac{1}{2}} \cos(n-1)x$ for $n \geq 2$, and $\lambda_n = (n-1)^2$. We take $\lambda_{j_o} = \lambda_1 = 0$ and $\lambda_{j_o+j_1} = \lambda_2 = 1$; so $\gamma = \frac{1}{2}$. In equation (2.17), we set G = 0, and take $f(x,s) = 2^{-1}s + \cos x$ for $s \geq 0, f(x,s) = -2^{-1}|s|^{\frac{1}{2}} + \cos x$ for $s \leq 0$. So equation (2.17) (as well as (2.9) in this case) becomes

$$-D_1 D_1 u = f(x,u) \tag{3.43}$$

and we shall show a solution of (3.43) exists in $H^1_{p,q,\varrho}$ via Theorem 5 but not via Theorem 4. To show this latter fact, we observe from (2.11) that $\mathcal{F}^+(x) = \mathcal{F}_+(x) = \frac{1}{2}$ and $\mathcal{F}^-(x) = \mathcal{F}_-(x) = 0$ $\forall x \in (0,\pi)$. It is clear that (2.12) holds in this case but on selecting w = -1, that (2.13) does not hold. Also, it is easy to see that f(x,s) meets (f-1), (f-2), (f-3), and (f-4) where $\gamma' = \frac{1}{4}$ and $f_0(x) = 1 + |\cos x|$ in (f-3). To show that a solution exists for the equation in (3.42) by Theorem 5, it remains to show that (2.16) holds.

To do this we are given a sequence $u_n = w_n + v_n$ where $u_n \in S_n$, $w_n = \hat{w}_n(1)\varphi_1$, $v_n(x) = \sum_{k=2}^n \hat{v}_n(k)\varphi_k(x)$ with $\hat{v}_n(k) = \langle v_n, \varphi_k \rangle_\varrho$, $\|u_n\|_\varrho^2 = |\hat{w}_n(1)|^2 + \sum_{k=2}^n |\hat{v}_n(k)|^2 \to \infty$, and $\|v_n\|_{p,q,\varrho} / \|u_n\|_{p,q,\varrho} \to 0$. Now

$$\|u_n\|^2_{p,q,\varrho} = |\hat{w}_n(1)|^2 + \sum_{k=2}^n \left[(k-1)^2 + 1\right] | \hat{v}_n(k) |^2 .$$

So we have that

$$|\hat{w}_n(1)| / \|u_n\|_{p,q,\varrho} \to 1.$$

Therefore we obtain that

$$|\hat{w}_n(1)| \to \infty \text{ and } \|v_n\|_{p,q,\varrho} / |\hat{w}_n(1)| \to 0. \tag{3.44}$$

From (3.44), we see that two cases present themselves.

I. $\exists$ a subsequence $\{n_j\}_{j=1}^\infty$ *s.t.* $\hat{w}_{n_j}(1) \to \infty$ or

II. $\hat{w}_n(1) \to -\infty$.

We deal with case I first and for ease of notation assume that $\hat{w}_n(1) \to \infty$. Now from the above, since $k^2 \leq 4(k-1)^2 + 4$, we see that $|v_n(x)| \leq \sum_{k=2}^{n} |\hat{v}_n(k)| \leq 2\pi \|v_n\|_{p,q,\varrho}$. Consequently, from (3.44)

$|v_n(x)| / \hat{w}_n(1) \to 0$ uniformly for $x \in \Omega$, and we conclude that in case I, there is an n_0 such that for $n \geq n_o$, $\hat{w}_n(1) > 0$ and

$$u_n(x) = \hat{w}_n(1)\,[\pi^{-\frac{1}{2}} + v_n(x)/\hat{w}_n(1)] \geq \frac{\pi^{-\frac{1}{2}}}{2}\hat{w}_n(1) \quad \forall x \in \Omega.$$

Consequently, for n $\geq n_0$, $f(x, u_n) = \frac{1}{2}$ u_n+ cos x, and

$$\langle f(x, u_n),\ w_n \rangle\, \varrho = \int_0^{\pi} f(x, u_n)\, w_n = \int_0^{\pi} \frac{1}{2} w_n^2 \to \infty$$

and (2.16) holds for case I.

For case II, $\hat{w}_n(1) \to -\infty$, and we conclude in the same manner as above that there is an n_0 such that for $n \geq n_o$, $\hat{w}_n(1) < 0$ and

$$u_n(x) = \hat{w}_n(1)\,[\pi^{-\frac{1}{2}} + v_n(x)/\hat{w}_n(1)] \leq \frac{\pi^{-\frac{1}{2}}}{2}\hat{w}_n(1) \quad \forall x \in \Omega.$$

Consequently, for n $\geq n_0$, $f(x, u_n) = -\frac{1}{2}|u_n|^{\frac{1}{2}} + \cos x$ and

$$\langle f(x, u_n), w_n \rangle_{\varrho} = \int_0^{\pi} -\frac{\pi^{-\frac{1}{2}}}{2}\hat{w}_n(1)\, |u_n|^{\frac{1}{2}}$$

But $|u_n|^{\frac{1}{2}} \geq |\hat{w}_n(1)|^{\frac{1}{2}}\, \pi^{-\frac{1}{4}}/\sqrt{2}$ for $n \geq n_0$. Consequently in case II,

$$\langle f(x, u_n), w_n \rangle_{\varrho} \geq \pi^{\frac{1}{4}}\, |\hat{w}_n(1)|^{3/2} / 2\sqrt{2} \quad \text{for } n \geq n_0.$$

Since $|\hat{w}_n(1)| \to \infty$, (2.16) holds for case II, and our example is complete.

To finish our discussion of Theorem 5 it remains to show by an example that Theorem 5 is in a certain sense a best possible result, i.e., if we replace γ' by γ in (f-3), then Theorem 5 is in general false. So we let

$$\text{(f-3)}' \quad |f(x,s)| \leq 2\gamma |s| + f_o(x), \quad f_o \in L^2_{\varrho}$$

where γ is defined in (2.10).

We take Ω, Γ, p, ϱ, q, and $L = Q$ exactly as in the last example. Also, we take $j_o = 1$, $j_1 = 1, \lambda_2 = 1$, $\gamma = \frac{1}{2}$, and $G = 0$. The only thing that changes from the last example is $f(x, s)$ which we now define as follows:

$$f(x,s) = s + cos\ x \quad \text{for} \quad s \in \mathbf{R} \text{ and } x \in \Omega.$$

Therefore we are dealing with equation (3.43) where f(x,s) is given by the last equality and $\lambda_{j_o} = \lambda_1 = 0$. Consequently, if we show $f(x,s)$ meets (2.16), then it will follow that all the conditions in the hypothesis of Theorem 5 are met with (f-3) replaced by (f-3)$'$.

Using the material from the last example, i.e., $u_n = w_n + v_n$ with $w_n = \hat{w}_n(1)\pi^{-\frac{1}{2}}$ and

$$v_n = \sum_{k=2}^{n} \hat{v}_n(k)\,\phi_k,$$

in conjunction with our new $f(x,s)$, we see that

$$\langle f(x,u_n), w_n\rangle_{\varrho} = \int_0^{\pi} (w_n + v_n + \cos x)w_n = \int_0^{\pi} w_n^2 = |\hat{w}_n(1)|^2 .$$

But as we saw in the last example, under the conditions given in (2.16), $|\hat{w}_n(1)|^2 \to \infty$. Consequently, the conclusion in (2.16) does indeed hold.

Now the equation in (2.17) is given by the equation in (3.43) and suppose that $u^* \in H^1_{p,q,\varrho}$ is a weak solution of this equation. Then, in particular, we would have that

$$\int_0^{\pi} D_1 u^* D_1 v = \int_0^{\pi} (u^* + \cos x)\, v \qquad \forall v \in H^1_{p,q,\varrho}$$

Choosing $v = \cos x$, in this last equation, we would then have that

$$-\int_0^{\pi} D_1 u^* \sin x = \int_0^{\pi} u^* \cos x \; + \pi/2.$$

But $u^* \in H^1_{p,q,\varrho}$ implies that the two integrals in this last equality are equal. Consequently, $0 = \pi/2$, giving a contradiction. Hence, no solution of (3.41) exists in $H^1_{p,q,\varrho}$, and our example showing that Theorem 5 does not hold in general if (f-3) is replaced by (f-3)$'$ is complete.

We close this section with an example which works for Theorem 12 and for which it is difficult to verify one of the conditions in the hypothesis of Theorem 11, namely that λ_1 is an L-pseudo-eigenvalue of Q.

For this example, we take L to be given by (3.10), and we have shown that the corresponding (Ω, Γ) is a V_L-region where $\Omega = (-1,1) \times (0,1)$ and $\Gamma = \{(t,1) : -1 \le t \le 1\}$. For Q, we take $A_i(x,t,\xi)$ to be given by (3.25) and $B_0(x) = 1$ where now p_1, p_2, q, and ϱ are defined in the two lines above (3.10). It follows from [KS,p.96] that Q meets (Q-1)-(Q-5) and that

$$\mathcal{Q}(u,v) - \mathcal{L}(u,v) = \int_{\Omega} \left[\sum_{i=1}^{2} p_i D_i u D_i v\right] \left[1 + |Du|^2\right]^{-\frac{1}{2}} . \qquad (3.45)$$

It is then clear that for $\|v\|_{p,q,\varrho} \leq 1$,

$$| \mathcal{Q}(u,v) - \mathcal{L}(u,v) | \leq \left[\int_\Omega p_1 + p_2 \right]^{\frac{1}{2}}.$$

Hence,Q is #-#-related to L and therefore #-related to L. Likewise, it is clear from (3.45) that $\mathcal{Q}(u,u) - \mathcal{L}(u,u) \geq 0$ for $u \in H^1_{p,q,\varrho}$. Consequently, (2.32) is valid, and this example indeed qualifies for Theorem 12.

1.4 Fundamental lemmas

The first lemma we prove is the following:

Lemma 1. *Assume that L is given by (1.6) and that the assumptions in (1.5) are valid. Assume furthermore that (Ω, Γ) is a V_L-region and that $g \in L^2_\varrho$. Set*

$$\hat{g}(n) = \langle g, \varphi_n \rangle_\varrho . \tag{4.1}$$

Then $g \in H^1_{p,q,\varrho}$ if and only if $\sum_{n=1}^\infty \lambda_n |\hat{g}(n)|^2 < \infty$. Furthermore if $g \in H^1_{p,q,\varrho}$, then $\mathcal{L}(g,g) = \sum_{n=1}^\infty \lambda_n |\hat{g}(n)|^2$.

We first of all observe from (1.4), (1.5), and (1.7) that

$$\mathcal{L}_1(u,v) = \mathcal{L}(u,v) + \langle u, v \rangle_\varrho \tag{4.2}$$

is a real inner product on $\mathrm{H}^1_{p,q,\varrho}$ and that there exists positive constants K_1 and K_2 such that

$$K_1 \|v\|^2_{p,q,\varrho} \leq \mathcal{L}_1(v,v) \leq K_2 \|v\|^2_{p,q,\varrho} \qquad \forall v \in H^1_{p,q,\varrho}. \tag{4.3}$$

Hence, $H^1_{p,q,\varrho}$ is a Hilbert space with respect to $\mathcal{L}_1(\cdot,\cdot)$.

Next, we see from (4.1), (4.2), and (V_L- 2) that

$$\mathcal{L}_1(v, \varphi_n/(\lambda_n+1)^{\frac{1}{2}}) = (\lambda_n+1)^{\frac{1}{2}} \hat{v}(n). \tag{4.4}$$

Consequently, we have from (V_L − 1) that

$$\{(\lambda_n+1)^{-\frac{1}{2}} \varphi_n\}_{n=1}^\infty \text{ is a } CONS \text{ on } H^1_{p,q,\varrho} \text{ wrt to } \mathcal{L}_1(\cdot,\cdot). \tag{4.5}$$

Hence it follows from (4.4) and Parseval's relationship that if $g \in H^1_{p,q,\varrho}$ then

$$\mathcal{L}_1(g,g) = \sum_{n=1}^{\infty} |\mathcal{L}_1(g,\varphi_n)|^2 / (\lambda_n + 1) = \sum_{n=1}^{\infty} (\lambda_n + 1) |\hat{g}(n)|^2 .$$

To complete the proof of the lemma, it remains to show that if $g \in L^2_\varrho$ and $\sum_{n=1}^{\infty}(\lambda_n + 1) |\hat{g}(n)|^2 < \infty$, then $g \in H^1_{p,q,\varrho}$. To accomplish this set,

$$h_n = \sum_{k=1}^{n} \hat{g}(k) \varphi_k = \sum_{k=1}^{n} (\lambda_k + 1)^{\frac{1}{2}} \hat{g}(k) \ \ \varphi_k \ \ |(\lambda_k + 1)|^{-\frac{1}{2}} .$$

Then, it follows from (4.5) that $\{h_n\}_{n=1}^{\infty}$ is a Cauchy sequence in $H^1_{p,q,\varrho}$ with respect to $\mathcal{L}_1(\cdot,\cdot)$ .Hence,

$$\exists h \in H^1_{p,q,\varrho} \text{ such that } \mathcal{L}_1(h - h_n, h - h_n) \to 0.$$

But then it follows from (4.3) that $\|h - h_n\|_\varrho \to 0$. However $\|h_n - g\|_\varrho \to 0$. Consequently $g = h$ and the proof of the lemma is complete.

Lemma 2. *Assume that L is given by (1.6), that the assumptions in (1.5) are valid, and that (Ω, Γ) is a V_L - region. Then $H^1_{p,q,\varrho}$ is compactly imbedded in L^2_ϱ.*

In Lemma 1, we showed that $\mathcal{L}_1(\cdot,\cdot)$ given by (4.2) is an equivalent inner product on $H^1_{p,q,\varrho}$ to $\langle\cdot,\cdot\rangle_{p,q,\varrho}$. So to prove the lemma, it follows from Lemma 1 that we have to show the following: Suppose $\{v_n\}_{n=1}^{\infty} \subset H^1_{p,q,\varrho}$ and

$$\mathcal{L}_1(v_n, v_n) \leq K \quad \forall n \text{ where } K \text{ is a positive constant.} \tag{4.6}$$

Then $\exists\, v \in H^1_{p,q,\varrho}$ and a subsequence $\{v_{n_m}\}_{m=1}^{\infty}$ such that

$$\lim_{m\to\infty} \|v_{n_m} - v\|_\varrho = 0 \tag{4.7}$$

From (4.6) and well-known properties of Hilbert spaces we see there exists $v \in H^1_{p,q,\varrho}$ and a subsequence $\{v_{n_m}\}_{m=1}^{\infty}$ such that $\mathcal{L}_1(v_{n_m}, w) \to \mathcal{L}_1(v, w)$ $\forall w \in H^1_{p,q,\varrho}$ and also that $\mathcal{L}_1(v,v) \leq K$. Consequently, we have from (4.1) and (4.4) that $\hat{v}_{n_m}(k) \to \hat{v}(k)$ $\forall k$ and, furthermore, that

$$\sum_{k=1}^{\infty} (\lambda_k + 1) |\hat{v}_{n_m}(k) - \hat{v}(k)|^2 \leq 4K \quad \forall m.$$

Now by $(V_L - 2)$, $\lambda_k \to \infty$. Therefore, given $\varepsilon > 0$, choose k_0 such that $4K/(\lambda_{k_0} + 1) < \varepsilon$. Then by $(V_L - 2)$ and the above inequality

$$\|v_{n_m} - v\|_{\varrho}^2 = \sum_{k=1}^{\infty} |\hat{v}_{n_m}(k) - \hat{v}(k)|^2$$

$$\leq \sum_{k=1}^{k_0} |\hat{v}_{n_m}(k) - \hat{v}(k)|^2 + ((\lambda_{k_0} + 1)^{-1} 4K)$$

Consequently, $\limsup_{m\to\infty} \|v_{n_m} - v\|_{\varrho}^2 \leq \varepsilon$. Since ε is arbitrary, (4.7) is established and the proof of the lemma is complete.

Next, we establish the following lemma:

Lemma 3. *Assume that the conditions in the hypothesis of Theorem 1 hold except for (2.3) and the relationship between the $\lambda_{j_o}-$ eigenfunctions of L and Q. Also, let S_n be the subspace of $H^1_{p,q,\varrho}$ spanned by $\varphi_1, ..., \varphi_n$. Then there exists $n_o > 1$, such that for every $n \geq n_o$, there is a $u_n \in S_n$ with the property that*

$$\mathcal{Q}(u_n, v) = (\lambda_{j_0} + n^{-1}) < u_n, v >_{\varrho} + G(u_n) < f, v >_{\varrho} \ \forall v \in S_n. \tag{4.8}$$

To prove the lemma, without loss of generality, we assume from the start that $j_o > 1$. (In case $j_o = 1$, a similar but easier proof prevails.) Also, we suppose that λ_{j_o} is an eigenvalue of L of order j_1. Hence

$$\varphi_{j_o}, \varphi_{j_o+1}, ..., \varphi_{j_o+j_1-1}$$

are all eigenfunctions corresponding to λ_{j_o}. We take n_o in the lemma to be the first positive integer greater than $j_o + j_1$ such that $\lambda_{j_o+\,j_1} - \lambda_{j_o} > 2/n_o$. So the fixed n that we deal with satisfies

$$n \geq j_o + j_1 + 1 \quad \text{and} \quad \lambda_{j_o+\,j_1} - \lambda_{j_o} > 2/n. \tag{4.9}$$

Next, with $\alpha = (\alpha_1, ..., \alpha_n) \in \mathbf{R}^n$, we set

$$u = \sum_{k=1}^{n} \alpha_k \ \varphi_k \ \text{ and } \quad \tilde{u} = \sum_{k=1}^{n} \delta_k \ \alpha_k \ \varphi_k \text{ where} \tag{4.10}$$

$$\delta_k = \{ {}^{-1}_{1} \ {}^{k=1,..., \ j_o+j_1-1}_{k=j_o+j_1,...,n.} \tag{4.11}$$

We define

$$F_k(\alpha) = \mathcal{Q}(u, \delta_k\varphi_k) - (\lambda_{j_o} + n^{-1}) < u, \delta_k\varphi_k >_{\varrho} - G(u) < f, \delta_k\varphi_k >_{\varrho} \tag{4.12}$$

for $k = 1..., n$.

It follows from (4.10) and (4.11) that

$$\sum_{k=1}^{n} F_k(\alpha)\,\alpha_k = \mathcal{L}(u,\tilde{u}) - \left(\lambda_{j_o} + n^{-1}\right) < u,\tilde{u} >_\varrho - G(u) < f,\tilde{u} >_\varrho \tag{4.13}$$

$$+Q(u,\tilde{u}) - \mathcal{L}(u,\tilde{u})$$

$$= \sum_{k=1}^{j_o+j_1-1} (\lambda_{j_o} + n^{-1} - \lambda_k)\alpha_k^2 + \sum_{k=j_o+j_1}^{n} \left(\lambda_k - \lambda_{j_o} - n^{-1}\right)\alpha_k^2$$

$$-G(u) < f,\tilde{u} >_\varrho + Q(u,\tilde{u}) - \mathcal{L}(u,\tilde{u}).$$

From $(V_L - 1)$, (4.10), and (4.11), we see that

$$\|u\|_\varrho^2 = \|\tilde{u}\|_\varrho^2 = |\alpha|^2 \tag{4.14}$$

Also, from the definition of j_1, we have that $\lambda_k \le \lambda_{j_o}$ for $k \le j_o + j_1 - 1$. Consequently, we obtain from (4.9) and (4.13) that

$$\sum_{k=1}^{n} F_k(\alpha)\,\alpha_k \ge |\alpha|^2/n - |G(u)|\ \|f\|_\varrho |\alpha| + Q(u,\tilde{u}) - \mathcal{L}(u,\tilde{u}) \tag{4.15}$$

Next, we observe from (4.2) and (4.10) that

$$\mathcal{L}_1(u,u) = \mathcal{L}_1(\tilde{u},\tilde{u}) = \sum_{k=1}^{n} (\lambda_k + 1)\,|\alpha_k|^2 \tag{4.16}$$

Since n is fixed and Q is #-related to L, we obtain from (1.11), (4.3), (4.14), and (4.16) that

$$\lim_{\|u\|_\varrho \to \infty} \left[Q(u,\tilde{u}) - \mathcal{L}(u,\tilde{u})\right]/\mathcal{L}_1(u,u) = 0$$

Since n is fixed and $\lambda_1 \ge 0$, it follows from (4.16) and this last fact that

$$\lim_{|\alpha| \to \infty} \left[Q(u,\tilde{u}) - \mathcal{L}(u,\tilde{u})\right]/|\alpha|^2 = 0. \tag{4.17}$$

Likewise, we see from (G-2) and (4.3) that

$$\lim_{\|u\|_\varrho \to \infty} |G(u)|\ \|f\|_\varrho/\ \left[\mathcal{L}_1(u,u)\right]^{\frac{1}{2}} = 0.$$

Hence, we obtain from (4.16) that

$$\lim_{|\alpha|\to\infty} |G(u)| \ \|f\|_{\varrho} |\alpha| / |\alpha|^2 = 0.$$

Since n is a fixed positive number, we conclude from (4.15), (4.17), and this last fact that

$$\exists\ s_o > 0 \text{ s.t. } \sum_{k=1}^{n} F_k(\alpha)\,\alpha_k > 0 \quad \text{for } |\alpha| \geq s_o. \tag{4.18}$$

Setting $F(\alpha) = [F_1(\alpha), ..., F_n(\alpha)]$, we observe from (4.12), (Q-1), (Q-2), (Q-5), and (G-1) that F is a continuous real map of $\mathbf{R}^n$ into $\mathbf{R}$. We consequently obtain from (4.18) (See [Ke, p. 219] or [Ni, p. 18].) that there exists $\alpha^* = (\alpha_1^*, ..., \alpha_n^*)$ such that $F_k(\alpha^*) = 0$ for $k = 1,\ldots,n$. In particular,$-F_k(\alpha^*) = 0$ for $k = 1, ..., j_o + j_1 - 1$. So taking $u_n = \sum_{k=1}^{n} \alpha_k^* \varphi_k$,we obtain from (4.11) and (4.12) that

$$Q(u_n, \varphi_k) = (\lambda_{j_o} + n^{-1}) < u_n, \varphi_k >_{\varrho} - G(u_n) < f, \varphi_k >_{\varrho}$$

for $k = 1, \ldots, n$. This fact when used with the definition of S_n, gives (4.8), and the proof of the lemma is complete.

We can obtain a lemma similar to Lemma 3 for the operator M, and we do so in the following.

Lemma 4. *Assume that the conditions in the hypothesis of Theorem 3 hold excluding (2.3) and the condition that λ_{j_o} be a near-eigenvalue of M. Let S_n be the subspace of $H^1_{p,q,\varrho}$ spanned by $\varphi_1, ..., \varphi_n$. Then there exists $n_o > 1$ such that for every $n \geq n_o$, there is a $u_n \in S_n$ with the property that*

$$\mathcal{M}(u_n, v) = \left(\lambda_{j_o} + n^{-1}\right) < u_n, v >_{\varrho} + G(u_n) < f, v >_{\varrho} \quad \forall v \in S_n. \tag{4.19}$$

To prove the lemma, we make the same assumptions concerning j_o that we did in the proof of Lemma 3. In particular, we assume that $j_o > 1$ and λ_{j_o} is an eigenvalue of L of multiplicity j_1. Also,for the fixed n we are dealing with, we assume (4.9), (4.10), (4.11), and define

$$F_k(\alpha) = \mathcal{M}(u, \delta_k\varphi_k) - \left(\lambda_{j_o} + n^{-1}\right) \langle u, \delta_k\varphi_k \rangle_{\varrho} - G(u) \langle f, \delta_k\varphi_k \rangle_{\varrho} \tag{4.20}$$

for k=1,..., n where $\alpha = (\alpha_1, ..., \alpha_n) \in \mathbf{R}^n$ and $\mathcal{M}$ is given by (2.5). It follows as in (4.13) that

$$\sum_{k=1}^{n} F_k(\alpha)\,\alpha_k = \sum_{k=1}^{j_o+j_1-1} \left(\lambda_{j_o} + n^{-1} - \lambda_k\right) \alpha_k^2 \tag{4.21}$$

$$+ \sum_{k=j_o+j_1}^{n} \left(\lambda_k - \lambda_{j_o} - n^{-1}\right) \alpha_k^2$$
$$- G(u) \quad < \quad f,\ \tilde{u} >_\varrho + \mathcal{M}(u,\tilde{u}) - \mathcal{L}(u,\tilde{u}) \quad .$$

From (4.9) and (4.21), we obtain (4.22) below in the same manner that we obtained (4.15) from (4.9) and (4.13).

$$\sum_{k=1}^{n} F_k(\alpha)\,\alpha_k \geq |\alpha|^2 / n - G(u)\,\|f\|_\varrho\,|\alpha| + \mathcal{M}(u,\tilde{u}) - \mathcal{L}(u,\tilde{u}). \tag{4.22}$$

Now from (1.7) and (2.5), we see that

$$\mathcal{M}(u,\tilde{u}) - \mathcal{L}(u,\tilde{u}) = \sum_{i,j=1}^{n} \int_\Omega p_i^{\frac{1}{2}} p_j^{\frac{1}{2}} \left[\sigma_i(u)\,\sigma_j(u)\text{-}1\right] a_{ij} D_j u D_i \tilde{u}$$
$$+ \left[\sigma_o(u) - 1\right] < a_o u, \tilde{u} >_q . \tag{4.23}$$

From (1.5)(i), it follows that $|< a_o u, \tilde{u} >_q| \leq K\,\|u\|_{p,q,\varrho}\,\|\tilde{u}\|_{p,q,\varrho}$, and therefore from (4.10) and (4.11), we obtain

$$\exists\ K_4 > 0 \text{ such that } |< a_o u, \tilde{u} >_q| \leq K_4\,|\alpha|^2 . \tag{4.24}$$

Similar reasoning using $(\sigma - 2)$ gives us that there exists $K_{ij} > 0$ such that

$$\left| \int_\Omega p_i^{\frac{1}{2}} p_j^{\frac{1}{2}} \left[\sigma_i(u)\,\sigma_j(u) - 1\right] a_{ij} D_j u D_i \tilde{u} \right| \leq K_{ij}\,|\sigma_i(u)\sigma_j(u) - 1|\,|\alpha|^2 \tag{4.25}$$

for i, j =1,..., N. Also from $(\sigma - 3)$ and (4.14), we obtain

$$\lim_{|\alpha|\to\infty} \sigma_i(u) = 1 \qquad \text{for } i = 0, 1, ..., N. \tag{4.26}$$

We consequently conclude from (4.23) -(4,26) that

$$\lim_{|\alpha|\to\infty} \left[\mathcal{M}(u,\tilde{u}) - \mathcal{L}(u,\tilde{u})\right] / |\alpha|^2 = 0. \tag{4.27}$$

Also, as in the proof of Lemma 3, using (G-2), we obtain

$$\lim_{|\alpha|\to\infty} |G(u)|\,\|f\|_\varrho / |\alpha| = 0. \tag{4.28}$$

From (4.22) with the help of (4.27) and (4.28), we obtain that

$$\exists s_o > 0\ s.t. \quad \sum_{k=1}^{n} F_k(\alpha)\,\alpha_k > 0 \quad \text{for } |\alpha| \geq s_o. \tag{4.29}$$

The conclusion of the proof the current lemma follows exactly from (4.29) as the analogous situation followed from (4.18) in the proof of Lemma 3, i.e., (4.29) implies (4.19) exactly in the same manner that (4.18) implied (4.8), and the proof of Lemma 4 is complete.

We next prove the following lemma which is useful in obtaining Theorem 5:

Lemma 5. *Assume all the conditions in the hypothesis of Theorem 4 excluding (2.12) and (2.13). Suppose, in addition, f satisfies (f-3) and (f-4) and that* $G \in \left[H^1_{p,q,\varrho}\right]'$. *Then* $\exists\ n_o > 1$ *such that for* $n \geq n_o$, *there is a* $u_n \in S_n$ *with the property that*

$$\mathcal{Q}\left(u_n, v\right) = (\lambda_{j_o} + \gamma' n^{-1}) < u_n, v >_\varrho \qquad (1.1)$$

$$+ \left(1 - n^{-1}\right) < f\left(\cdot, u_n\right), v >_\varrho - G\left(v\right) \qquad \forall v \in S_n$$

where γ' *is given in (f-3).*

We first of all observe that (f-3) and (f-4) together give (2.15) where $f_o \in L^2_\varrho$ and $\gamma' > 0$.

To prove the lemma, we assume $j_o > 1$. (Theorem similar but easier proof prevails in case $j_o = 1$.) We also assume that λ_{j_o} is of multiplicity j_1. We take u and ũ as in (4.10) and δ_k as in (4.11). We shall also assume $n_o^{-1} < \gamma'$ and that $n_o \geq j_o + j_1 + 1$. Next, we define for $\alpha = (\alpha_1, ..., \alpha_n) \in \mathbf{R}^n$ and $n \geq n_o$ (with n understood to be fixed)

$$\begin{aligned} F_k\left(\alpha\right) &= \mathcal{Q}\left(u, \delta_k \varphi_k\right) - (\lambda_{j_o} + \gamma' n^{-1}) < u, \delta_k \phi_k >_\varrho \qquad (4.31) \\ -\left(1 - n^{-1}\right) &< f\left(\cdot, u\right), \delta_k \phi_k >_\varrho - G\left(\delta_k \phi_k\right) \qquad \text{for } k = 1, ..., n. \end{aligned}$$

Also, we set

$$\mathrm{I}(\alpha) = \mathcal{L}\left(u, \tilde{u}\right) - \left(\lambda_{j_o} + \gamma'\right) < u, \tilde{u} >_\varrho \qquad (4.32)$$

$$-(1 - n^{-1}) < f(\cdot, u) - \gamma' u, \tilde{u} >_\varrho - G\left(\tilde{u}\right)$$

and

$$\mathrm{II}\left(\alpha\right) = Q\left(u, \tilde{u}\right) - \mathcal{L}\left(u, \tilde{u}\right). \qquad (4.33)$$

Then, we see from the above that

$$\sum_{k=1}^{n} F_k\left(\alpha\right) \alpha_k = \mathrm{I}\left(\alpha\right) + \mathrm{II}\left(\alpha\right). \qquad (4.34)$$

Now, as (4.13),

$$\mathcal{L}(u,\tilde{u}) - (\lambda_{j_o}+\gamma') < u,\tilde{u} >_\varrho$$

$$= \sum_{k=1}^{j_o+j_1-1} (\lambda_{j_o}+\gamma' - \lambda_k)\alpha_k^2 + \sum_{k=j_o+j_1}^{n} (\lambda_k - \lambda_{j_o} - \gamma')\alpha_k^2$$

Since $2\gamma' < \lambda_{j_o+j_1} - \lambda_{j_o}$, we obtain from the above computation that

$$\mathcal{L}(u,\tilde{u}) - \left(\lambda_{j_o}+\gamma'\right) < u,\tilde{u} >_\varrho \geq \gamma' \left|\alpha\right|^2.$$

Hence, it follows from (2.15) and (4.32) that

$$\mathrm{I}(\alpha) \geq \gamma' \left|\alpha\right|^2 - \left(1-n^{-1}\right)\left[\left(\gamma' \left|\alpha\right| + \left\|f_o\right\|_\varrho\right)\left|\alpha\right|\right] - K\left|\alpha\right| \tag{4.35}$$

where K is a constant independent of α.

Since Q is #-related to L, it also follows (as we have shown in Lemma 3) that

$$\lim_{|\alpha|\to\infty} \mathrm{II}(\alpha) / \left|\alpha\right|^2 = 0 \tag{4.36}$$

We consequently conclude from (4.34)-(4.36) that there exists $s_o > 0$ such that

$$\sum_{k=1}^{n} F_k(\alpha)\,\alpha_k > \gamma' \left|\alpha\right|^2/2n > 0 \quad \text{for } \left|\alpha\right| \geq s_o.$$

This last established fact is the same as (4.18) in the proof of Lemma 3. The rest of the proof of Lemma 5 proceeds in a very similar manner to that of Lemma 3, and we leave the details to the reader.

Lemma 6 *Let Ω, p_i, ϱ, q, G, L, Q, and Γ be as in the hypothesis of Theorem 6. Suppose, also, that f satisfies (f-1), (f-6), and (f-7) and that λ_{j_o} is an eigenvalue of L of multiplicity j_1. Then $\exists\ n_o > j_o$ such that for $n \geq n_o$, there is a $u_n \in S_n$ with the property that*

$$Q(u_n,v) = \left(\lambda_{j_o} - n^{-1}\right) < u_n, v >_\varrho + (1+n^{-1}) < f(\cdot,u_n), v >_\varrho - G(v) \tag{4.37}$$

$\forall v \in S_n$.

To prove this lemma, we assume (as in the previous lemmas) that $j_o > 1$ and λ_{j_o} is of multiplicity j_1. The case $\lambda_{j_o} = \lambda_1$ follows in a similar manner. Also, we set

$$n_o = \text{first integer} \geq max\left[j_o + j_1 + 1, \quad 2/(\lambda_{j_o} - \lambda_{j_o-1})\right]. \tag{4.38}$$

For $\alpha = (\alpha_1, ..., \alpha_n) \in \mathbf{R}^n$, we take u and ũ as in (4.10) but now we define δ_k differently then in (4.11), namely

$$\delta_k = \begin{cases} -1 & k=1,\ldots,j_o-1 \\ 1 & k=j_o,\ldots,\ n \end{cases} \tag{4.39}$$

We define for $n \geq n_o$ and $\alpha \in \mathbf{R}^n$,

$$F_k(\alpha) = \mathcal{Q}(u, \delta_k\varphi_k) - \left(\lambda_{j_o} - n^{-1}\right) < u, \delta_k\varphi_k >_\varrho \tag{4.40}$$

$$+ \left(1 + n^{-1}\right) < f(\cdot, u), \delta_k\varphi_k >_\varrho - G(\delta_k\varphi_k) \quad \text{for } k = 1, \ldots, n.$$

Also, we set

$$I(\alpha) = \mathcal{L}(u, \tilde{u}) - \left(\lambda_{j_o} - n^{-1}\right) < u, \tilde{u} >_\varrho \tag{4.41}$$

$$+ \left(1 + n^{-1}\right) < f(\cdot, u), \tilde{u} >_\varrho - G(\tilde{u})$$

and

$$II(\alpha) = \mathcal{Q}(u, \tilde{u}) - \mathcal{L}(u, \tilde{u}) \ . \tag{4.42}$$

Then

$$\sum_{k=1}^{n} F_k(\alpha)\, \alpha_k = I(\alpha) + II(\alpha), \tag{4.43}$$

and it follows from (4.10) and (4.39) that

$$\mathcal{L}(u, \tilde{u}) - \left(\lambda_{j_o} - n^{-1}\right) < u, \tilde{u} >_\varrho =$$

$$\sum_{k=1}^{j_o-1} \left[\lambda_{j_o} - \left(\lambda_k + n^{-1}\right)\right] \alpha_k^2 + \sum_{k=j_o}^{n} \left(\lambda_k + n^{-1} - \lambda_{j_o}\right) \alpha_k^2$$

From (4.38), we see that $(\lambda_{j_o} - \lambda_k) \geq 2/n$ for $k = 1, ..., j_o - 1$. Hence, we infer from this last computation that

$$\mathcal{L}(u,\tilde{u}) - (\lambda_{j_o} - n^{-1}) < u,\tilde{u} >_\varrho \geq n^{-1} |\alpha|^2 .$$

But then it follows from (4.41), on choosing $\varepsilon = (4n)^{-1}$ in (f-7), and from this last computation that

$$I(\alpha) \geq n^{-1} | \alpha |^2 - (2n)^{-1} \|u\|_\varrho \|\tilde{u}\|_\varrho - \| b_{(4n)^{-1}} \|_\varrho \ \|\tilde{u}\|_\varrho - K \|\tilde{u}\|_{p,q,\varrho}$$

where K is a constant that depends upon G and $b_{(4n)^{-1}}(x)$ is defined in (f-7). It follows from (4.10) that $\|u\|_\varrho = \|\tilde{u}\|_\varrho = |\alpha|$. Hence, we obtain from (4.3) and (4.10) that there is an $s_o > 0$ such that

$$I(\alpha) \geq (4n)^{-1} |\alpha|^2 \quad \text{for } |\alpha| \geq s_o. \tag{4.44}$$

Now, Q is #-related to L. So it follows from (4.3), (4.10), and (4.42) that there is an $s \geq s_o$ such that

$$|II(\alpha)| \leq (8n)^{-1} |\alpha|^2 \quad \text{for } |\alpha| \geq s_1.$$

We conclude from (4.43) and (4.44) that

$$\sum_{k=1}^{n} F_k(\alpha)\alpha_k \geq (8n)^{-1} |\alpha| \quad \text{for } |\alpha| \geq s_1.$$

This last established inequality is the same as (4.18) in Lemma 3. The rest of the proof of Lemma 6 proceeds in a similar manner to that of Lemma 3 (the difference in the definitions of δ_k has to be taken into consideration), and we leave the details to the reader.

We next establish a continuous imbedding lemma for Simple V_L-regions which will be very useful in the proof of the compact imbedding Theorem 9. The lemma that we shall prove is the following.

Lemma 7. *Let Lu be given by (2.24) and suppose that* (Ω,Γ) *is a Simple* V_L *-region. Then* $H^1_{p,q,\varrho}(\Omega,\Gamma)$ *is continuously imbedded in* $L^\theta_\varrho(\Omega)$ *for every* θ *satisfying* $2 < \theta < 2N/(N-1)$, *i.e.,* $\exists K_\theta > 0$ *such that*

$$\left\{ \int_\Omega |u|^\theta \varrho dx \right\}^{1/\theta} \leq K_\theta \|u\|_{p,q,\varrho} \qquad \forall u \in H^1_{p,q,\varrho}. \tag{4.45}$$

It is clear to establish the proof of the lemma we need only consider the case when N$\geq$ 2 and where u in (4.45) with $u \in H^1_{p,q,\varrho}(\Omega,\Gamma)$ is replaced by v with $v \in C^1_{p,q,\varrho}(\Omega,\Gamma)$ (See(1.2).).

Now $\Omega = \Omega_1 \times \cdots \times \Omega_N$, and for fixed $x_2, ..., x_N$, $v(x_1, x_2, ..., x_N)$ as a function of x_1 is in $C^2(\Omega_1)$. Hence, it follows from (2.25) (iv) and (2.26) that

$|v(x_1, x_2, ..., x_N)| \leq h_1(x_1)\, g_1^{\frac{1}{2}}(x_2, ..., x_N)$ where
$g_1(x_2, ..., x_N) =$

$$\int_{\Omega_1} \left[|D_1 v(x_1, x_2, ..., x_N)|^2 p_1^*(x_1) + |v(x_1, x_2, ..., x_N)|^2 \varrho_1^*(x_1) \right] dx_1.$$

Applying the same kind of reasoning to x_i for fixed $x_1, ..., x_{i-1}, x_{i+1}, \ldots, x_N$, we conclude that

$$|v(x)|^N = \prod_{i=1}^{N} h_i(x_i) \prod_{i=1}^{N} g_i^{\frac{1}{2}}(\check{x}_i) \qquad \text{for } x \in \Omega \tag{4.46}$$

where

$$\check{x}_i = (x_1, ..., x_{i-1}, x_{i+1}, ..., x_N) \tag{4.47}$$

and

$$g_i(\check{x}_i) = \int_{\Omega_i} \left[|D_i v(x)|^2 p_i^*(x_i) + |v(x)|^2 \varrho_i^*(x_i) \right] dx_i \tag{4.48}$$

With

$$2 < \theta < 2N/(N-1) \qquad \text{where } N \geq 2, \tag{4.49}$$

we see from (4.46) that

$$|v(x)|\,\theta \leq \prod_{i=2}^{N} |\, h_i(x_i)\,|^{\theta/N} |g_1(\check{x}_1)|^{\theta/2N} |h_1(x_1)|^{\theta/N} \prod_{i=2}^{N} |\, g_i(\check{x}_i)\,|^{\theta/2N}$$

Applying the generalized version of Holder's inequality on Ω_1, [Zy, p. 18], with $\sum_{i=1}^{n} r_i^{-1} = 1$ where $r_1 = \beta$ and $r_2 = ... = r_N = \gamma$ with $\gamma = 2N/\theta$ and $\beta = 2N/[2N - (N-1)\theta]$ to the N terms in the above inequality which are functions of x_1, we obtain

$$\int_{\Omega_1} |v(x)|^\theta \varrho_1^*(x_1)\, dx_1$$

$$\leq \prod_{i=2}^{N} |\, h_i(x_i)\,|^{\theta/N} |g_1(\check{x}_1)|^{\theta/2N} [\int_{\Omega_1} |\, h_1(x_1)\,|^{\beta\theta/N} \varrho_1^*(x_1)\, dx_1\,]^{1/\beta}$$

$$\prod_{i=2}^{N} \left[\int_{\Omega_1} g_i(\check{x}_i)\, \varrho_1^*(x_1)\, dx_1 \right]^{1/\gamma}$$

Applying this procedure once again on Ω_2 to the N terms which are functions of x_2, we obtain

$$\int_{\Omega_2}\int_{\Omega_1} |v(x)|^{\theta}\, \varrho_1^*(x_1)\, \varrho_2^*(x_2)\, dx_1 dx_2$$

$$\leq \prod_{i=3}^{N} | h_i(x_i) |^{\theta/N} \left[\int_{\Omega_1} | g_2(\check{x}_2)\, \varrho_1^*(x_1)\, dx_1\right]^{1/\gamma}$$
$$\prod_{i=1}^{2} \left[\int_{\Omega_i} | h_i(x_i) |^{\beta\theta/N}\, \varrho_i^*(x_i)\, dx_i\right]^{1/\beta}$$
$$\left[\int_{\Omega_2} | g_1(\check{x}_1)\, \varrho_2^*(x_2)\, dx_2\right]^{1/\gamma}$$

$$\prod_{i=3}^{N} [\int_{\Omega_2}\int_{\Omega_1} g_i(\check{x}_i)\, \varrho_1^*(x_1)\, \varrho_2^*(x_2)\, dx_1 dx\,]^{1/\gamma}$$

Iterating this procedure a total of N-times and recalling that $\varrho(x) = \varrho_1^*(x_1) ... \varrho_N^*(x_N,)$ we see from (4.46) that

$$\int_{\Omega} |v(x)|^{\theta}\, \varrho(x)\, dx \leq$$

$$\prod_{i=1}^{N} \left[\int_{\Omega_i} | h_i(x_i) |^{\beta\theta/N}\, \varrho_i^*(x_i)\, dx_i\right]^{1/\beta} \left[\prod_{i=1}^{N} J_i\right]^{1/\gamma} \tag{4.50}$$

where

$J_i =$

$$\int_{\Omega_N} ... \int_{\Omega i+1}\int_{\Omega_{i-1}} ... \int_{\Omega_1} g_i(\check{x}_i)\, \varrho_1^*(x_1) ... \varrho_{i-1}^*(x_{i-1})\, \varrho_{i+1}^*(x_{i+1}) ... \varrho_N^*(x_N)\, d\check{x}_i \tag{4.51}$$

Now from (2.25) (iv),we see that $h_i(x_i) \in L_{\varrho i*}^{\beta\theta/N}(\Omega_i)$. Also, we recall that $\gamma = 2N/\theta$. Hence, we obtain from (4.50) that there is a constant K_θ such that

$$[\int_{\Omega} |v(x)|^{\theta}\, \varrho(x)\, dx\,]^{1/\theta} \leq K_\theta \left[\prod_{i=1}^{N} J_i\right]^{1/2N} \tag{4.52}$$

We next recall that

$$p_i(x) = \varrho_1^*(x_1)...\varrho_{i-1}^*(x_{i-1})\, p_i^*(x_i)\, \varrho_{i+1}^*(x_{i+1})...\varrho_N^*(x_N).$$

We consequently see from (4.48) and (4.51) that

$$J_i = \int_\Omega \left[|D_i v(x)|^2 p_i(x) + |v(x)|^2 \varrho(x) \right] dx$$

Therefore $J_i \leq \|v\|_{p,q,\varrho}^2$ for $i = 1,...,N$. But, then

$$\prod_{i=1}^N J_i \leq \|v\|_{p,q,\varrho}^{2N}$$

and (4.45) with v replacing u follows from (4.52) and this last inequality. The proof of Lemma 7 is complete.

Lemma 8 *Let all the conditions in the hypotheses of Theorem 10 hold. Then for $n \geq 2$, there is a $u_n \in S_n$ with the property that*

$$\mathcal{Q}(u_n, v) = (\lambda_1 - n^{-1}) < u_n, v >_\varrho + \int_\Omega f(x, u_n) v \varrho - G(v) \tag{4.53}$$

$\forall v \in S_n.$

To establish this lemma, we fix n and set

$$u = \sum_{k=1}^n \alpha_k \varphi_k$$

where $\alpha = (\alpha_1, ..., \alpha_n) \in \mathbf{R}^n$ and observe that

$$\mathcal{L}(u,u) = \sum_{k=1}^n \lambda_k |\alpha_k|^2 \text{ and } \|u\|_\varrho^2 = \sum_{k=1}^n \alpha_k^2$$

From (1.3) and (2.24), we have that $\mathcal{L}(u,u) + \|u\|_\varrho^2 = \|u\|_{p,q,\varrho}^2$. Consequently we conclude from $(\mathrm{V}_L - 2)$ that

$$\|u\|_{p,q,\varrho}^2 \leq (\lambda_n + 1) \|u\|_\varrho^2 \qquad \forall u \in S_n. \tag{4.54}$$

Next for $m \geq 2$, a positive integer, we put

$$f_m(x,s) = \begin{matrix} f(x,m) & \text{for } m \leq s \\ f(x,s) & \text{for } -m \leq s \leq m \end{matrix}$$

$$f(x,-m) \qquad \text{for } s \leq -m$$

and claim that the following fact obtains:

$$\exists u_m^* \in S_n \text{ such that} \tag{4.55}$$

$$\mathcal{Q}(u_m^*, v) = \left(\lambda_1 - n^{-1}\right) < u_m^*, v >_\varrho + \textstyle\int_\Omega f_m(x, u_m^*) v\varrho - G(v)$$

$$\forall v \in S_n.$$

By (f-8) and the above

$$|f_m(x,s)| \leq h_o(x) + K|m|^{\theta-1} \qquad \forall s \in \mathbf{R} \text{ and a.e. } x \in \Omega \tag{4.56}$$

where $h_o \in L_\varrho^{\theta^*}(\Omega)$ and $\theta^* = \theta/(\theta-1)$. Therefore, $|f_m(x,u)| \in L_\varrho^{\theta^*}(\Omega)$ $\forall u \in S_n$ and $\int_\Omega f_m(x,u)v\varrho$ is well defined for $v \in S_n$ by Lemma 7. To establish (4.55), we set

$$F_k(\alpha) = \mathcal{Q}(u, \varphi_k) - \left(\lambda_1 - n^{-1}\right) < u, \varphi_k >_\varrho \tag{4.57}$$

$$-\int_\Omega f_m(x,u)\,\varphi_k \varrho - G(\varphi_k) \qquad \text{for } k = 1, ..., n.$$

and observe that

$$\sum_{k=1}^n F_k(\alpha)\,\alpha_k = \mathrm{I}(\alpha) + \mathrm{II}(\alpha). \tag{4.58}$$

where

$$\mathrm{I}(\alpha) = \mathcal{L}(u,u) - \left(\lambda_1 - n^{-1}\right) < u, u > \varrho \tag{4.59}$$

and

$$\mathrm{II}(\alpha) = -\int_\Omega f_m(x,u)u\varrho - G(u) + Q(u,u) - \mathcal{L}(u,u) \tag{4.60}$$

Now from the first paragraph of this proof and (4.59),

$$I(\alpha) = \sum_{k=1}^n (\lambda_k - \lambda_1)\,\alpha_k^2 + |\alpha|^2/n.$$

On the other hand, it follows from Lemma 7, (4.54), (4.56), (4.60), and the fact that Q is #-related to L that

$$\lim_{|\alpha|\to\infty} |\,\mathrm{II}(\alpha)|\,/\,(\alpha)^2 = 0.$$

We conclude from (4.58) that $\exists s_o > 0$ such that

$$\sum_{k=1}^{n} F_k(\alpha)\alpha_k \geq |\alpha|^2 /2n \qquad \text{for } |\alpha| \geq s_o.$$

Therefore, as in many of the previous lemmas, $\exists\ \gamma = (\gamma_1, ..., \gamma_n)$ such that $F_k(\gamma) = 0$ for $k = 1, ..., n$. We set $u_m^* = \sum_{k=1}^{n} \gamma_k\ \varphi_k$ and see from (4.57) that (4.55) does indeed hold.

Proceeding with the proof of the lemma we now have a sequence of functions $\{\mathrm{u}_m^*\}_{m=1}^{\infty} \subset S_n$ such that (4.55) does hold. We next observe from the definition of $f_m(x,s)$ and (2.28) that for $m \geq 2$,

$$sf_m(x,s) \leq \tilde{h}_o(x)\,|s| \qquad \forall s \in \mathbf{R} \text{ and } a.e.\ x \in \Omega. \tag{4.61}$$

where $\tilde{h}_o(x) \in L_\varrho^{\theta^*}(\Omega)$.

We claim

$$\exists K_1 > 0\ s.t. \quad \|u_m^*\|_\varrho \leq K_1 \quad \forall\ m \geq 2. \tag{4.62}$$

Suppose that (4.62) is false. Then there is a subsequence (which for ease of notation, we take to be the full sequence) such that

$$\lim_{|m|\to\infty} \|u_m^*\|_\varrho = \infty. \tag{4.63}$$

To see that (4.63) leads to a contradiction, we insert u_m^* in place of v in (4.55) and obtain

$$\mathcal{L}\left(u_m^*, u_m^*\right) - \lambda_1 < u_m^*, u_m^* >_\varrho + n^{-1} < u_m^*, u_m^* >_\varrho$$

$$= \textstyle\int_\Omega f_m\left(\cdot, u_m^*\right) u_m^* \varrho\ -\ G\left(u_m^*\right)$$

$$+\mathcal{L}\left(u_m^*, u_m^*\right) - \mathcal{Q}(u_m^*, u_m^*)$$

Now, as before, the left-hand side of this last equality is

$$\sum_{k=1}^{n} \left(\lambda_k - \lambda_1\right) \left|\hat{u}_m^*\left(k\right)\right|^2 + \|u_m^*\|_\varrho^2 /n.$$

So we see from (4.61) and Holder's inequality applied to the right-hand side that

$$\|u_m^*\|_\varrho^2 /n \leq \left\|\tilde{h}_o\right\|_{L_\varrho^{\theta^*}} \|\ u_m^*\ \|_{L_\varrho^{\theta}} + \left|G\left(u_m^*\right)\right| + \mathcal{Q}(u_m^*, u_m^*) - \mathcal{L}\left(u_m^*, u_m^*\right)$$

Dividing both sides of this last inequality by $\|u_m^*\|_\varrho^2$ and leaving m$\to\infty$, we obtain from (4.45), (4.54), (4.63), and the fact that Q is #-related to L that $n^{-1} \leq 0$. But n is a positive integer. So we have arrived at a contradiction. Therefore, (4.62) does indeed hold.

Now $u_m^* \in S_n$ for m ≥ 2. Therefore (4.62) and (4.54) along with the rest of the first paragraph of this proof implies there is a subsequence $\{u_{m_j}^*\}_{j=1}^\infty$ and a $u_n \in S_n$ such that

$$\lim_{j\to\infty} \left\| u_{m_j}^* - u_n \right\|_{p,q,\varrho} = 0, \quad \lim_{j\to\infty} u_{m_j}^* (x) = u_n (x) \text{ for a.e. } x \in \Omega,$$

$$\text{and } \lim_{j\to\infty} D_i u_{m_j}^* (x) = D_i u_n (x) \text{ for a.e. } x \in \Omega,\ i = 1, ..., N.$$

As a consequence it is easy to see from (Q-1) -(Q-5) that $\mathcal{Q}(u_m^*, v) \to \mathcal{Q}(u_n, v)$ as $j \to \infty$ $\forall v \in S_n$. On the other-hand from Lemma 7 (the continuous imbedding theorem), it follows that $\lim_{j\to\infty} \int_\Omega \left| u_{m_j}^* - u_n \right|^\theta \varrho = 0$. Consequently, $\exists W \in L_\varrho^\theta (\Omega)$ and a subsequence $\left\{ u_{m_{j_k}} \right\}_{k=1}^\infty$ such that $| u_{m_{j_k}}^* (x) | \leq W (x)$ a.e. in Ω $\forall k$. We also have

$$\lim_{k\to\infty} f_{m_{j_k}} (x, u_{m_{j_k}}^*) = f(x, u_n) \text{ a.e. in } \Omega.$$

We therefore conclude from (f-8), Holder's inequality, and the Lebesgue dominated convergence theorem that

$$\lim_{k\to\infty} \int_\Omega f_{m_{j_k}} (x, u_{m_{j_k}}^*) v\varrho = \int_\Omega f(x, u_n) v\varrho \quad \forall v \in S_n.$$

Hence replacing u_m^* by $u_{m_{j_k}}^*$ in (4.55) and passing to the limit on both sides of the equal sign as $k \to \infty$, we consequently obtain that

$$\mathcal{Q}(u_n, v) = \left(\lambda_1 - n^{-1}\right) < u_n, v >_\varrho + \textstyle\int_\Omega f(x, u_n) v\varrho - G (v) \quad \forall v \in S_n, \text{ and}$$

the proof of the lemma is complete.

1.5 Proofs of Theorem 1 and Corollary 2

To prove the necessary condition of Theorem 1, assume $u^* \in H_{p,q,\varrho}^1$ and that (2.4) holds for every $v \in H_{p,q,\varrho}^1$. Let φ be a $\lambda_{j_0}-$ eigenfunction of L. Then it is a $H_{p,q,\varrho}^1 - \lambda_{j_0}-$ eigenfunction of Q by hypothesis. Consequently

$\mathcal{Q}(u^*, \varphi) = \lambda_{j_o} < u^*, \varphi >_\varrho$. Therefore, on taking φ for v in (2.4), one obtains that $G(u^*) < f, \varphi >_\varrho = 0$. By (G-3), $G(u^*) \neq 0$. Hence $< f, \varphi >_\varrho = 0$, and the necessary condition of Theorem 1 is established.

To obtain the sufficiency condition of Theorem 1, we assume without loss in generality that $j_o > 1$. (In case $j_o = 1$, a similar but easier proof prevails.) Also, we shall suppose that λ_{j_0} is an eigenvalue of L of order j_1. Hence

$$\varphi_{j_o}, \varphi_{j_o+1}, ..., \varphi_{j_o+j_1-1} \text{ are orthonormal eigenfunctions of L corresponding to the eigenvalue } \lambda_{j_o}. \tag{5.1}$$

Next, we invoke Lemma 3 and obtain for every integer $n \geq n_o$, a function $u_n \in S_n$ with the property that

$$\mathcal{Q}(u_n, v) = \left(\lambda_{j_o} + n^{-1}\right) < u_n, v >_\varrho + G(u_n) < f, v >_\varrho \quad \forall v \in S_n. \tag{5.2}$$

Now,

$$u_n = \sum_{k=1}^{n} \hat{u}_n(k)\, \varphi_k.$$

We write

$$\begin{aligned} (i) \quad u_n &= u_{n1} + u_{n2} + u_{n3} \quad \text{where} \\ (ii) \; u_{n1} &= \sum_{k=1}^{j_o-1} \hat{u}_n(k)\, \varphi_k, \qquad u_{n2} = \sum_{k=j_o}^{j_o+j_1-1} \hat{u}_n(k)\, \varphi_k, \\ u_{n3} &= \sum_{k=\, j_o+j_1}^{n} \hat{u}_n(k)\, \varphi_k. \end{aligned} \tag{5.3}$$

From the hypothesis of the theorem, we have that every $\lambda_{j0}-$ eigenfunction of L is a $H^1_{p,q,\varrho}-\lambda_{j0}-$eigenfunction of Q. Therefore we have that $\mathcal{Q}(u_n, \varphi_k) = \lambda_{j_o} < u_n, \varphi_k >_\varrho$ for $k = j_o, ..., j_o + j_1 - 1$. Consequently, we see from (5.3)(ii) that

$$\mathcal{Q}(u_n, u_{n2}) = \lambda_{j_o} < u_n, u_{n2} >_\varrho = \lambda_{j_o} < u_{n2}, u_{n2} >_\varrho \tag{5.4}$$

Also from (2.3) and (5.3)(ii), we see that $< f, u_{n2} >_\varrho = 0$. Therefore, using u_{n2} for v in (5.2), we obtain from this last fact and (5.4) that

$$\lambda_{j_o} < u_{n2}, u_{n2} >_\varrho = \left(\lambda_{j_o} + n^{-1}\right) < u_{n2}, u_{n2} >_\varrho$$

But then $n^{-1} < u_{n2}, u_{n2} >_\varrho = 0$, and we conclude that

$$u_{n2} = 0 \tag{5.5}$$

We claim

$$\exists\, K > 0 \text{ s.t. } \|u_n\|_{p,q,\varrho} \leq K \qquad \text{for } n \geq n_o. \tag{5.6}$$

Suppose to the contrary that (5.6) does not hold. For ease of notation and without loss of generality, we assume

$$\lim_{n\to\infty} \|u_n\|_{p,q,\varrho} = \infty \tag{5.7}$$

We will show this assumption gives a contradiction.

To accomplish this, using (5.3) and (5.5), we define

$$\tilde{u}_n = -\, u_{n1} + u_{n3} \tag{5.8}$$

and put $\tilde{u}_n$ in place of v in (5.2) to obtain from $(V_L - 1)$ that

$$\mathcal{Q}(u_n, \tilde{u}_n) = (\lambda_{jo} + n^{-1})(-\|u_{n1}\|_\varrho^2 + \|u_{n3}\|_\varrho^2) + G(u_n) < f, \tilde{u}_n >_\varrho \tag{5.9}$$

Next we observe from $(V_L - 2)$, (5.3) and (5.8) that

$$\mathcal{L}(u_n, \tilde{u}_n) = -\sum_{j=1}^{j_o-1} \lambda_j \left|\hat{u}_n(j)\right|^2 + \sum_{j=j_o+j_1}^{n} \lambda_j \left|\hat{u}_n(j)\right|^2 \tag{5.10}$$

Consequently, we obtain from (5.9) and (5.10) that

$$\begin{aligned}\sum_{j=1}^{j_o-1}(\lambda_{j_o} + n^{-1} - \lambda_j)\left|\hat{u}_n(j)\right|^2 &+ \sum_{j=j_o+j_1}^{n} (\lambda_j - \lambda_{j_o} - n^{-1})\left|\hat{u}_n(j)\right|^2 \\ &= \mathcal{L}(u_n, \tilde{u}_n) - \mathcal{Q}(u_n, \tilde{u}_n) + G(u_n) < f, \tilde{u}_n >_\varrho\end{aligned} \tag{5.11}$$

Now from $(V_L - 2)$, we see

$$\begin{aligned}&(i)\ \exists \gamma_1 > 0 \quad s.t.\ (\lambda_{j_o} - \lambda_j) \geq \gamma_1 [\lambda_j + 1] \text{ for } j = 1, ..., j_o - 1, \text{ and} \\ &(ii)\ \exists \gamma_2 > 0 \text{ and } \exists n_1 \text{ with } n_1 \geq n_o \ s.t. \\ &\qquad (\lambda_j - \lambda_{j_o} - n^{-1}) \geq \gamma_2(\lambda_j + 1) \text{ for } j \geq j_o + j_1 \text{ and } n \geq n_1.\end{aligned} \tag{5.12}$$

Set

$$\gamma_3 = min\ (\gamma_1, \gamma_2) > 0. \tag{5.13}$$

We then obtain from (5.11)-(5.13) and (5.5) that

$$\gamma_3 \sum_{j=1}^{n} (\lambda_j + 1)\left|\hat{u}_n(j)\right|^2 \leq \mathcal{L}(u_n, \tilde{u}_n) - \mathcal{Q}(u_n, \tilde{u}_n) + G(u_n) < f, \tilde{u}_n >_\varrho. \tag{5.14}$$

Next, we observe from (5.3) that

$$\mathcal{L}(u_n, u_n) = \sum_{j=1}^{n} \lambda_j \left|\hat{u}_n(j)\right|^2 \text{ and that } \|u_n\|_\varrho^2 = \sum_{j=1}^{n} \left|\hat{u}_n(j)\right|^2.$$

Consequently, with $\mathcal{L}_1(\cdot,\cdot)$ defined by (4.2), we obtain from(5.14) that

$$\gamma_3^2 \, \mathcal{L}_1(u_n, u_n) \leq \mathcal{L}(u_n, \tilde{u}_n) - \mathcal{Q}(u_n, \tilde{u}_n) + G(u_n) < f, \tilde{u}_n >_\varrho \tag{5.15}$$

Now, $\mathcal{L}_1(u_n, u_n) = \mathcal{L}_1(\tilde{u}_n, \tilde{u}_n)$. Therefore, it follows from the fact Q is #-related to L, (4.3), (5.7) that

$$\lim_{n\to\infty} \left[\mathcal{L}(u_n, \tilde{u}_n) - \mathcal{Q}(u_n, \tilde{u}_n)\right] / \mathcal{L}_1(u_n, u_n) = 0 \tag{5.16}$$

Also, $|< f, u_n >_\varrho| \leq \|f\|_\varrho \|u_n\|_{p,q,\varrho}$. Therefore, it likewise follows from (G-2), (4.3), and (5.7) that

$$\lim_{n\to\infty} G(u_n) < f, \tilde{u}_n >_\varrho / \mathcal{L}_1(u_n, u_n) = 0 \tag{5.17}$$

Hence dividing both sides of (5.15) by $\mathcal{L}_1(u_n, u_n)$ and taking the limit as $n \to \infty$, we obtain that from (5.16) and (5.17) that

$$\gamma_3 \leq 0. \tag{5.18}$$

But from (5.13) we see that $\gamma_3 > 0$. Consequently, using (5.7) we have arrived at a contradiction. So (5.7) is false, and (5.6) is indeed true.

Since $H^1_{p,q,\varrho}$ is a separable Hilbert space, we see from (5.6), (G-1), and Lemma 2 that there exists a subsequence (which for ease of notation we take to be the full sequence) and a function $u^* \in \mathrm{H}^1_{p,q,\varrho}$ with the following properties:

$$\lim_{n\to\infty} \|u_n - u^*\|_\varrho = 0, \tag{5.19}$$

$$\exists w \in L^2_\varrho \ s.t. \left|u_n(x)\right| \leq w(x) \text{ for } a.e.\, x \in \Omega, n = 1, 2, ..., \tag{5.20}$$

$$\lim_{n\to\infty} u_n(x) = u^*(x) \text{ for } a.e.\ x \in \Omega, \tag{5.21}$$

$$\lim_{n\to\infty} < D_i u_n, v >_{p_i} = < D_i u^*, v >_{p_i} \ \forall v \in L^2_{p_i} \text{ and } i = 1, ..., N. \tag{5.22}$$

$$\lim_{n\to\infty} < B_o u_n, v >_q = < B_o u^*, v >_q \ \forall v \in L_q. \tag{5.23}$$

$$\lim_{n\to\infty} G(u_n) = G(u^*). \tag{5.24}$$

Next, we propose to show the following: there exists a subsequence $\{u_{n_j}\}_{j=1}^{\infty}$ such that

$$\lim_{j\to\infty} Du_{n_j}(x) = Du^*(x) \quad \text{for } a.e.\ x \in \Omega. \tag{5.25}$$

To establish (5.25) it is sufficient to establish the following two facts:
(1) There exists a subsequence $\{u_{n_j}\}_{j=1}^{\infty}$ such that

$$\lim_{j\to\infty} \sum_{i=1}^{N} p_i^{\frac{1}{2}}(x) \{ [\, A_i(x, u_{n_j}, Du_{n_j}) - A_i(x, u_{n_j}, Du^*) \,] \tag{5.26}$$

$$[D_i u_{n_j}(x) - D_i u^*(x)] \,\} = 0 \text{ for } a.\ e.\ x \in \Omega$$

(2) With $\{u_{n_j}\}_{j=1}^{\infty}$ designating the same subsequence as in (5.26),

$$\{|\, Du_{n_j}(x)\, |\}_{j=1}^{\infty} \text{ is pointwise bounded for } a.e.\ x \in \Omega \tag{5.27}$$

To see that (5.26) and (5.27) together imply (5.25), let Ω_1 be the subset of Ω for which (Q-1), (Q-4), (5.21), (5.26), and (5.27) hold simultaneously. Consequently,

$$meas\,(\Omega - \Omega_1) = 0 \tag{5.28}$$

Suppose there exists $x_o \in \Omega_1$ such that (5.25) does not hold. Hence by (5.27) there exists a further sequence $\left\{Du_{n_{j_k}}(x_o)\right\}_{k=1}^{\infty}$ and a $\xi^{\#} \in \mathbf{R}^N$ with

$$Du^*(x_o) \neq \xi^{\#} \tag{5.29}$$

such that $\lim_{n\to\infty} Du_{n_{j_k}}(x_o) = \xi^{\#}$. Therefore from (5.21)

$$\lim_{k\to\infty} \sum_{i=1}^{N} p_i^{\frac{1}{2}}(x_o) \{ [\, A_i\left(x_o, u_{n_{j_k}}, Du_{n_{j_k}}\right) - A_i\left(x_o, u_{n_{j_k}}, Du^*\right) \,] \tag{5.30}$$

$$\left[D_i u_{n_{j_k}}(x) - D_i u^*(x)\right] \}$$

$$= \sum_{i=1}^{N} p_i^{\frac{1}{2}}(x_o) \{ [\, A_i\left(x_o, u^*(x_o), \xi^{\#}\right) - A_i(x_o, u^*(x_o), Du^*(x_o)) \,]$$

$$[\, \xi_i^{\#} - D_i u^*(x_o) \,] \}$$

From (Q-4) and (5.29), we see that the right- hand side of (5.30) is strictly positive. Hence the limit on the left-hand side of (5.30) is strictly positive. But $x_o \in \Omega_1$, and by the definition of Ω_1 and (5.26), this limit is zero. We have therefore arrived at a contradiction. Hence the equality in (5.25) holds at every point in Ω_1, and consequently by (5.28) almost everywhere in Ω. So statement (5.25) is fully established.

It remains to show that (5.26) and (5.27) hold. To establish (5.26), we observe from (Q-2), (5.20) and (5.21) that

$$\lim_{n\to\infty}\int_\Omega \sum_{i=1}^N |A_i(x,u_n,Du^*) - A_i(x,u^*,Du^*)|^2 = 0 \tag{5.31}$$

Also, we have from (5.22) that

$$\lim_{n\to\infty}\int_\Omega A_i(x,u^*,Du^*)\,[D_iu_n - D_iu^*]\,p_i^{\frac{1}{2}} = 0 \text{ for i=1,...,N.} \tag{5.32}$$

Hence, it follows from (5.6), (5.31), and (5.32) that

$$\lim_{n\to\infty}\int_\Omega \sum_{i=1}^N A_i(x,u_n,Du^*)\,[D_iu_n - D_iu^*]\,p_i^{\frac{1}{2}} = 0 \tag{5.33}$$

Likewise, it follows from (5.23) and (Q-5) that

$$\lim_{n\to\infty}\int_\Omega B_0(x)u^*\,(u_n - u^*)\,q = 0 \tag{5.34}$$

Now if we can show that

$$\lim_{n\to\infty} \mathcal{Q}\,(u_n, u_n - u^*) = 0, \tag{5.35}$$

then it will follow from (1.10), (5.33), and (5.34) that

$$\lim_{n\to\infty}\int_\Omega \{\sum_{i=1}^N [A_i(x,u_n,Du_n) - A_i(x,u_n,Du^*)] \tag{5.36}$$

$$[D_iu_n - D_iu^*]\,p_i^{\frac{1}{2}} + B_0(x)\,(u_n - u^*)\,(u_n - u^*)\,q\,\} = 0$$

This last fact, in turn, implies (5.26). For from (Q-4) and (Q-5), we have that the integrand in (5.36) is nonnegative almost everywhere in Ω. Hence the integrand converges in $L^1(\Omega)$ to zero. But then by [Rud, p.70], a subsequence of the integrand converges almost everywhere in Ω to zero. By (5.21), we have that $B_o(x)\left|u_n(x)-u^*(x)\right|^2$ converges to zero almost everywhere in Ω. We conclude that (5.26) is indeed true. So to establish (5.26) it remains to establish (5.35).

To show that (5.35) is indeed true, we observe that $u^* \in \mathrm{H}^1_{p,q,\varrho}$, that

$$P_n u^* = \sum_{k=1}^{n} \hat{u}^*(k)\,\varphi_k \text{ is in } S_n, \tag{5.37}$$

and from Lemma 1 and (4.3) that

$$\lim_{n\to\infty} \|P_n u^* - u^*\|_{p,q,\varrho} = 0 \tag{5.38}$$

Now

$$\mathcal{Q}(u_n, P_n u^* - u^*) = \sum_{i=1}^{N} \int_\Omega A_i(x, u_n, Du_n)(P_n u^* - u^*) p_i^{\frac{1}{2}}$$
$$+ < B_o u_n, P_n u^* - u^* >_q .$$

and it follows from (Q-2), (Q-5), (5.6), and (5.38) that

$$\lim_{n\to\infty} \mathcal{Q}(u_n, P_n u^* - u^*) = 0.$$

Therefore (5.35) will follow if we show

$$\lim_{n\to\infty} \mathcal{Q}(u_n, u_n - P_n u^*) = 0 \tag{5.39}$$

From (5.2) and (5.37), we see that

$$\mathcal{Q}(u_n, u_n - P_n u^*) = \tag{5.40}$$
$$(\lambda_{j_o} + n^{-1}) < u_n, u_n - P_n u^* >_\varrho + G(u_n) < f, u_n - P_n u^* >_\varrho$$

But $< u_n, u_n - P_n u^* >_\varrho = < u_n, u_n - u^* >_\varrho + < u_n, u^* - P_n u^* >_\varrho$and we see (5.19) and (5.38) that the first term on the right-hand side of (5.40) tends to zero as $n \to \infty$. Likewise, from (5.19) and (5.38), the fact that $G(u_n)$ is uniformly bounded (from (5.6) and (G-2)), and the fact that $f \in L\varrho^2$, we see that the second term on the right-side of (5.40) tends to zero as $n \to \infty$. Hence the right-side of (5.40) tends to zero. We conclude that (5.39) is indeed true. Therefore (5.35) is correct and hence (5.26) is established.

To establish (5.27), we let Ω_2 be the set where simultaneously

$$u^*(x), |Du^*(x)|, h_i^*(x), u_{n_j}(x), A_i\left(x, u_{n_j}(x), Du_{n_j}(x)\right)$$

$$\text{and } A_i\left(x, u_{n_j}(x), u^*(x)\right) \text{ are finite-valued}$$

for $i = 1, ..., N$ and $j = 1, 2, ...,$ where (Q-2) and (Q-3) hold, and also where the limits in (5.21) and (5.26) exist. Then

$$meas\ (\Omega - \Omega_2) = 0. \tag{5.41}$$

Suppose, to the contrary, that $\left\{| Du_{n_j}(x) |\right\}_{j=1}^{\infty}$ is not pointwise bounded for $x \in \Omega_2$. Then there exists $x_o \in \Omega_2$ and a subsequence $\left\{| Du_{n_{j_k}}(x_o) |\right\}_{k=1}^{\infty}$ such that

$$\lim_{k\to\infty} \left|Du_{n_{j_k}}(x_o)\right| = \infty \tag{5.42}$$

Set

$$c_3 = min\left[p_1(x_o), ..., p_N(x_o)\right], \quad \therefore\ c_3 > 0. \tag{5.43}$$

From (Q-3), we have

$$\sum_{i=1}^{N} p_i^{\frac{1}{2}}(x_o)\, A_i\left(x_o, u_{n_{j_k}}, Du_{n_{j_k}}\right) D_i u_{n_{j_k}}(x_o) \geq c_2 c_3 \left|Du_{n_{j_k}}(x_o)\right|^2 \tag{5.44}$$

Also,

$$\sum_{i=1}^{N} p_i^{\frac{1}{2}}(x_o)\, A_i\left(x_o, u_{n_{j_k}}, Du_{n_{j_k}}\right) D_i u_{n_{j_k}}(x_o) \tag{5.45}$$

$$= \sum_{i=1}^{N} p_i^{\frac{1}{2}}(x_o)\, A_i\left(x_o, u_{n_{j_k}}, D_i u_{n_{j_k}}\right) D_i u^*(x_o)$$

$$+ \sum_{i=1}^{N} p_i^{\frac{1}{2}}(x_o)\, A_i\left(x_o, u_{n_{j_k}}, Du^*\right) \left[D_i u_{n_{j_k}}(x_o) - D_i u^*(x_o)\right]$$

$$+ \sum_{i=1}^{N} p_i^{\frac{1}{2}}(x_o)\left[A_i\left(x_o, u_{n_{j_k}}, Du_{n_{j_k}}\right) - A_i\left(x_o, u_{n_{j_k}}, Du^*\right)\right]$$

$$\left[D_i u_{n_{j_k}}(x_o) - D_i u^*(x_o)\right]$$

We divide both sides of (5.44) by $\left|Du_{n_{j_k}}(x_o)\right|^{3/2}$ and pass to the limit as $k\to\infty$. We conclude from (5.45), (Q-2), (5.21), (5.26), and the definition of Ω_2 that

$$0 \geq c_2 c_3 \lim_{k\to\infty} \left|Du_{n_{j_k}}(x_o)\right|^{\frac{1}{2}}$$

From (Q-3) and (5.43) we see that $c_2c_3 > 0$. Hence, we have that

$$\lim_{k\to\infty} \left| Du_{n_{j_k}}(x_o) \right| = 0.$$

But this contradicts (5.42). Consequently $\left\{ \left| Du_{n_{j_k}}(x_o) \right| \right\}_{j=1}^{\infty}$ is pointwise bounded at every $x \in \Omega_2$ This fact in conjunction with (5.41) establishes (5.27) and this, in turn, in conjunction with (5.26) establishes (5.25) as we have already shown.

We now have that (5.19)-(5.25) hold and also that (5.2) and (5.6) hold. We proceed with the proof and let $v_J \in S_J$ where $J \geq n_o$ is a fixed but arbitrary positive integer. From (5.6) and (Q-2) we have that the sequence

$$\left\{ \int_\Omega \left| A_i\left(x, u_{n_j}, Du_{n_j}\right) - A_i(x, u^*, Du^*) \right|^2 \right\}_{j=1}^{\infty}$$

is a uniformly bounded sequence for $i = 1, ..., N$. Also, it follows from (5.21) and (5.25) that

$$\lim_{j\to\infty} \left| A_i\left(x, u_{n_j}, Du_{n_j}\right) - A_i(x, u^*, Du^*) \right| = 0 \quad \text{a.e. in } \Omega.$$

Now $D_i v_J \in L^2_{p_i}$. Consequently, we obtain from Schwarz's inequality and from Egoroff's theorem (see [Rud, p. 75]) applied to the positive measure $\mu_i(E) = \int_E p_i(x)\, dx$ that for $i = 1, \ldots, N$,

$$\lim_{j\to\infty} \int_\Omega \left[A_i\left(x, u_{n_j}, Du_{n_j}\right) - A_i(x, u^*, Du^*) \right] D_i v_J \, p_i^{\frac{1}{2}} = 0.$$

We conclude from (1.10), (5.2), (5.19), (5.24), and this last established fact that

$$\mathcal{Q}(u^*, v_J) = \lambda_{j_o} < u^*, v_J >_\varrho + G(u^*) < f, v_J >_\varrho . \tag{5.46}$$

(We note for future reference that (5.24) makes use of (G-1).)

Next, given $v \in \mathrm{H}^1_{p,q,\varrho}$, define $\mathrm{P}_J v$ as in (5.37) and observe as before from Lemma 1 and (4.3) that $\lim_{J\to\infty} \| P_J v - v \|_{p,q,\varrho} = 0$. Consequently, it follows that

$$\lim_{J\to\infty} \mathcal{Q}(u^*, P_J v) = \mathcal{Q}(u^*, v),\ \lim_{J\to\infty} < u^*, P_J v >_\varrho = < u^*, v >_\varrho,$$

and

$$\lim_{J\to\infty} < f, P_J v >_\varrho = < f, v >_\varrho$$

On the other hand from the equation in the preceding paragraph involving v_J, we have that

$$\mathcal{Q}(u^*, P_J v) = \lambda_{j_o} < u^*, P_J v >_\varrho + G(u^*) < f, P_J v >_\varrho.$$

Passing to the limit as J$\to \infty$ on both sides of this last equality and using the above established facts, we obtain

$$\mathcal{Q}(u^*, v) = \lambda_{j_o} < u^*, v >_\varrho + G(u^*) < f, v >_\varrho \; \forall v \in H^1_{p,q,\varrho};$$

and the proof of Theorem 1 is complete.

To prove Corollary 2, we assume that $q = 0$ in the above proof of Theorem 1 and observe that the only place (outside of Lemma 3 where we worked in a finite-dimensional space) where we used the fact that $(G-1)$ holds, i.e., where G is weakly sequentially continuous and (5.24) holds, is near the very end of the proof in the establishing of (5.46), and where we already knew that (5.2), (5.6), (5.19)-(5.23), and (5.25) holds. If using (5.2), (5.6), and (5.19)-(5.23), we can also show that

$$\left\{ \mid D_i u_n^2 \mid p_i \right\}_{n=1}^{\infty} \text{ is absolutely equi-integrable in } \mathrm{L}^1(\Omega) \tag{5.47}$$

for $i = 1, ..., N$, then the whole proof of the Corollary will follow in a manner similar to the proof given for Theorem 1. For then (5.25) in conjunction with (5.47) and Egoroff's theorem implies that $\left\| u_{n_j} - u^* \right\|_{p,\varrho} \to 0$ as $j \to \infty$. Hence, $(G-1)'$ replaces (5.24) and we obtain that $\lim_{j\to\infty} G(u_{n_j}) = G(u^*)$. Therefore, (5.46) holds, and the proof of the Corollary proceeds as before along the same lines as Theorem 1. (We need $\mu(\Omega) < \infty$ where μ is Lebesgue in order to apply Egoroff's Theorem.)

We have to establish (5.47), which explicitly means:

$$\text{given } \varepsilon > 0, \; \exists \vartheta \; s.t. \; \mu(E) < \vartheta \implies \int_E |D_i u_n|^2 p_i < \varepsilon \;\; \forall n$$

and for $i = 1, ..., N$. To prove (5.47), we have at our disposal (5.2), (5.6), (5.19)-(5.23), along with (G-2). A check of the proof Theorem 1 shows that these items are sufficient to establish (5.36) with $q = 0$ which we now also have at our disposal.

So we set

$$Y_n(x) = \sum_{i=1}^{N} [A_i(x, u_n, Du_n) - A_i(x, u_n, Du^*)][D_i u_n - D_i u^*] p_i^{\frac{1}{2}} \tag{5.48}$$

and observe from (Q-4) that $Y_n(x) \geq 0$ a.e. in Ω. Hence it follows from (5.36) and the fact that $q = 0$ that $Y_n(x) \to 0$ in $L^1(\Omega)$. But then it follows that

$$\{Y_n\}_{n=1}^{\infty} \text{ is absolutely equi-integrable in } L^1(\Omega). \tag{5.49}$$

Also, it is clear from (5.6), (5.20), (Q-2), and Schwarz's inequality that both sequences

$$\left\{A_i(x, u_n, Du^*)(D_iu_n - D_iu^*)p_i^{\frac{1}{2}}\right\}_{n=1}^{\infty}$$

$$\text{and } \left\{A_i(x, u_n, Du_n) D_iu^*p_i^{\frac{1}{2}}\right\}_{n=1}^{\infty}$$

are absolutely equi-integrable in $L^1(\Omega)$ for $i = 1, ..., N$.

Next, we observe from (5.48) that

$$\sum_{i=1}^{N} A_i(x,, u_n, Du_n) D_iu_np_i^{\frac{1}{2}} = \sum_{i=1}^{N} A_i(x, u_n, Du_n) D_iu^*p_i^{\frac{1}{2}}$$

$$+\sum_{i=1}^{N} A_i(x, u_n, Du^*)[D_iu_n - D_iu^*]p_i^{\frac{1}{2}} + Y_n(x),$$

and we conclude from (5.49) that the sequence obtained from the left-hand side of the above equality is absolutely equi-integrable in $L^1(\Omega)$. However by (Q-3)

$$\sum_{i=1}^{N} A_i(x,, u_n, Du_n) D_iu_np_i^{\frac{1}{2}} \geq c_2 \sum_{i=1}^{N} p_i(x) |D_iu_n|^2$$

for a.e. $x \in \Omega$ where $c_2 > 0$. Consequently, the sequence obtained from the right-hand side of this last stated inequality is absolutely equin-tegrable, (5.47) is established, and the proof of Corollary 2 is complete.

1.6 Proof of Theorem 3

Without loss in generality we can assume from the start the $j_o > 1$. (a similar but easier proof prevails in case $j_o = 1$.). We suppose as in the proof of the sufficiency condition of Theorem 1 that λ_{j_o}, a near-eigenvalue of M, is also an eigenvalue of L of multiplicity j_1. Hence we assume that (5.1) holds.

Next, we invoke Lemma 4 and obtain for every integer $n \geq n_o$, a function $u_n \in S_n$ with the property that

$$\mathcal{M}(u_n, v) = \left(\lambda_{j_o} + n^{-1}\right) < u_n, v >_\varrho + G(u_n) < f, v >_\varrho \quad \forall v \in S_{n.} \tag{6.1}$$

We write u_n as in (5.3) (i) and (ii) and obtain from (2.6) and the fact that λ_{j_o} is a near-eigenvalue of M that

$$\mathcal{M}(u_n, u_{n2}) = \sigma^{\ddagger}(u_n) \lambda_{j_o} < u_n, u_{n2} >_\varrho \tag{6.2}$$

We also see from (5.3) that $< u_n, u_{n2} >_\varrho = \|u_{n2}\|_\varrho^2$ and from (2.3) that $< f, u_{n2} >= 0$.. Setting $v = u_{n2}$ in (6.1), we therefore obtain from (6.2) that

$$[\sigma^{\ddagger}(u_n) - 1] \lambda_{j0} \|u_{n2}\|_\varrho^2 = n^{-1} \|u_{n2}\|_\varrho^2$$

Since $\sigma^{\ddagger}$ satisfies $(\sigma - 4)$, $\sigma^{\ddagger}(u_n) - 1 \leq 0$. Hence, from this last equality, we see that $\|u_{n2}\|_\varrho \leq 0$. Consequently we conclude from (5.3) that

$$(i) \quad u_n = u_{n1} + u_{n3} \quad \text{where} \tag{6.3}$$

$$(ii) \quad u_{n1} = \sum_{j=1}^{j_o-1} \hat{u}_n(j) \phi_j \text{ and } u_{n3} = \sum_{j=j_o+j_1}^{n} \hat{u}_n(j) \phi_j.$$

We claim

$$\exists K > 0 \ s.t \ \|u_n\|_{p,q,\varrho} \leq K \quad \text{for } n \geq n_o. \tag{6.4}$$

Suppose to the contrary, that (6.4) does not hold. For ease of notation and without loss of generality, we can then assume that

$$\lim_{n\to\infty} \|u_n\|_{p,q,\varrho} = \infty \tag{6.5}$$

We will show this assumption gives a contradiction.

First of all we claim, (6.5) implies that

$$\lim_{n\to\infty} \|u_n\|_\varrho = \infty \tag{6.6}$$

Suppose once again to the contrary, that

$$\exists K > 0 \ s.t \ \|u_n\|_\varrho \leq K_1 \quad \forall n \geq n_o. \tag{6.7}$$

It follows from (1.5), (2.5), and $(\sigma - 2)$ that

$$\mathcal{M}(u_n, u_n) \geq \eta_o c_o \sum_{i=1}^{N} \|D_i u_n\|_{p_i}^2 + \varepsilon_o \eta_o \|u_n\|_q^2$$

where ε_o, c_o, η_o are positive constants. Therefore putting $v = u_n$ in (6.1), we obtain from this last inequality that

$$\exists c_1 > 0 \text{ s.t.} \tag{6.8}$$

$$c_1 \left[\sum_{i=1}^{N} \|D_i u_n\|_{p_i}^2 + \|u_n\|_q^2 \right] \leq (\lambda_{j_0} + n^{-1}) \|u_n\|_\varrho^2 + |G(u_n)| \, \|f\|_\varrho \|u_n\|_\varrho$$

for n $\geq n_o$. Now from (1.4) and (G-2), we have that for every n,

$$|G(u_n)| \leq 2 \left[\sum_{i=1}^{N} \|D_i u_n\|_{p_i} + \|u_n\|_\varrho + \|u_n\|_q \right] + K_2 \tag{6.9}$$

where K_2 is a constant. But (1.4), (6.5), and (6.7) imply that

$$\lim_{n\to\infty} E_n = \infty \text{ where } E_n = \sum_{i=1}^{N} \|D_i u_n\|_{p_i}^2 + \|u_n\|_q^2 . \tag{6.10}$$

Hence, we have from (6.7) and (6.9) that

$$\lim_{n\to\infty} |G(u_n)| / E_n = 0. \tag{6.11}$$

Consequently, if we divide both sides of (6.8) by E_n and pass to the limit as $n \to \infty$, we obtain from (6.7), (6.10), and (6.11) that $c_1 \leq 0$. But $c_1 > 0$, and we have arrived at a contradiction. Therefore (6.6) is indeed true.

Next, using (6.3) (i) and (ii), we set

$$\tilde{u}_n = - u_{n1} + u_{n3} \tag{6.12}$$

and put $\tilde{u}_n$ in (6.1) in place of v to obtain from $(V_L - 1)$ that

$$\mathcal{M}(u_n, \tilde{u}_n) = (\lambda_{j0} + n^{-1})[- \|u_{n1}\|_\varrho^2 + \|u_{n3}\|_\varrho^2] + G(u_n) < f, \tilde{u}_n >_\varrho \tag{6.13}$$

Also, we observe from $(V_L - 2)$, (6.3), and (6.12) that (5.10) holds. Hence, we obtain from (6.13) that

$$\sum_{j=1}^{j_o-1} (\lambda_{j_0} + n^{-1} - \lambda_j) |\hat{u}_n(j)|^2 + \sum_{j=j_o+j_1}^{n} (\lambda_j - \lambda_{j_0} - n^{-1}) |\hat{u}_n(j)|^2 \tag{6.14}$$

$$= \mathcal{L}(u_n, \tilde{u}_n) - \mathcal{M}(u_n, \tilde{u}_n) + G(u_n) < f, \tilde{u}_n >_\varrho$$

Now, once again (5.12) (i) and (ii) are valid. Defining γ_3 as in (5.13), we consequently obtain from (6.14) that

$$\gamma_3 \sum_{j=1}^{n} (\lambda_j+1) \left|\hat{u}_n(j)\right|^2 \leq \mathcal{L}(u_n,\tilde{u}_n) - \mathcal{M}(u_n,\tilde{u}_n) + G(u_n) < f, \tilde{u}_n >_\varrho \quad (6.15)$$

for $n \geq n_1$. where $\gamma_3 > 0$.

Next, we see that

$$\mathcal{M}(u_n,\tilde{u}_n) - \mathcal{L}(u_n,\tilde{u}_n) =$$

$$\sum_{i,j=1}^{N} \int_{\hat{\Omega}} a_{ij}(x) \left[\sigma_i(u_n)\sigma_j(u_n)\text{-}1\right] p_i^{\frac{1}{2}} p_j^{\frac{1}{2}} D_j u_n D_i \tilde{u}_n \quad (6.16)$$

$$+ \int_{\hat{\Omega}} a_0(x) \left[\sigma_0(u_n)\text{-}1\right] q\, u_n \tilde{u}_n$$

Also, we observe from (6.6) and $(\sigma - 3)$ that $\lim_{n\to\infty} \sigma_i(u_n)\sigma_j(u_n) = 1$. Hence, it follows from (1.5),(6.16), and Schwarz's inequality that

$$\text{Given } \varepsilon > 0, \exists n_\varepsilon \text{ s.t. for } n \geq n_\varepsilon \quad (6.17)$$

$$\left|\mathcal{L}(u_n,\tilde{u}_n) - \mathcal{M}(u_n,\tilde{u}_n)\right| \leq K_3\varepsilon \; \|u_n\|_{p,q,\varrho} \; \|\tilde{u}_n\|_{p,q,\varrho}$$

where K_3 is a positive constant independent of ε. From Lemma 1, (6.3), and (6.12), we see from (4.2) that $\mathcal{L}_1(u_n,u_n) = \mathcal{L}_1(\tilde{u}_n,\tilde{u}_n)$. Hence it follows from (4.3) that

$$\exists K_4 > 0 \text{ s.t. } \|\tilde{u}_n\|_{p,q,\varrho} \leq K_4[\mathcal{L}_1(u_n,u_n)]^{\frac{1}{2}}. \quad (6.18)$$

We consequently obtain from (6.17) and (6.18) that

$$\lim_{n\to\infty} \left|\mathcal{L}(u_n,\tilde{u}_n) - \mathcal{M}(u_n,\tilde{u}_n)\right| / \mathcal{L}_1(u_n,u_n) = 0. \quad (6.19)$$

In a similar manner, we obtain from (G-2) that

$$\lim_{n\to\infty} \left|G(u_n) < f, u_n >_\varrho\right| / \mathcal{L}_1(u_n,u_n) = 0. \quad (6.20)$$

From (4.2) and (6.3), we also obtain

$$\mathcal{L}_1(u_n,u_n) = \sum_{j=1}^{n} (\lambda_j+1) \left|\hat{u}_n(j)\right|^2 \quad (6.21)$$

Dividing both sides of (6.15) by $\mathcal{L}_1(u_n, u_n)$ and leaving $n \to \infty$, we conclude from (6.17) - (6.21) that $\gamma_3 \leq 0$. But $\gamma_3 > 0$. Hence, we have arrived at a contradiction. So (6.5) is false and (6.4) is indeed true.

Next, we observe exactly as in the proof of Theorem 1 that (6.4) implies there exists a subsequence (we take the full sequence for ease of notation) and a function $u^* \in H^1_{p,q,\varrho}$ such that (5.19)-(5.24) holds where B_o is replaced by a_o in (5.23) and furthermore such that

$$\lim_{n\to\infty} \sigma_i(u_n) = \sigma_i(u^*) \qquad i = 0, ..., N \tag{6.22}$$

where we have used the fact that σ_i meets $(\sigma - 1)$.

Next, we return to (6.1) and assume that $v_J \in S_J$ where $J \geq n_{o.}$ Now, it follows from (2.5), (5.19), (5.22), and (6.22) that

$$\lim_{n\to\infty} \mathcal{M}(u_n, v_J) = \mathcal{M}(u^*, v_J) . \tag{6.23}$$

Hence, replacing v by v_J in (6.1) and taking the limit as $n \to \infty$ on both sides of the equation, we obtain from (5.19), (5.24), and (6.23) that

$$\mathcal{M}(u^*, v_J) = \lambda_{j0} < u^*, v_J >_\varrho +G(u^*) < f, v_J >_\varrho \tag{6.24}$$

for $v_J \in S_J$, $J \geq n_{o.}$ Next, given $v \in H^1_{p,q,\varrho}$, define $P_J v$ as in (5.37). Then $\mathrm{P}_J v \in S_J$ and $\|P_J v - v\|_{p,q,\varrho} \to 0$ as $J \to \infty$ by Lemma 1 and (4.3). It is easy to see, therefore, from (1.5) and (2.5) that $\mathcal{M}(u^*, P_J v) \to \mathcal{M}(u^*, v)$. Also, $< u^*, P_J v >_\varrho \to < u^*, v >_\varrho$ and $< f, v_J >_\varrho \to < f, v >_\varrho$. Hence it follows from (6.24) that

$$\mathcal{M}(u^*, v) = \lambda_{j0} < u^*, v >_\varrho +G(u^*) < f, v >_\varrho \ \forall v \in H^1_{p,q,\varrho}.$$

Thus (2.8) is established, and the proof of Theorem 3 is complete.

1.7 Proof of Theorem 5

Without loss in generality we assume from the start that $j_o > 1$ (a similar but easier proof prevails in case $j_o = 1$) and that γ' in (f-3) is strictly positive. Hence the conditions in the hypothesis of Lemma 5 hold, and we invoke Lemma 5 and obtain a sequence of functions

$$\{u_n\}_{n=1}^{\infty} \text{ in } H^1_{p,q,\varrho} \text{ such that}$$

$$\mathcal{Q}(u_n, v) = (\lambda_{j0} + \gamma' n^{-1}) < u_n, v >_\varrho + (1 - n^{-1}) < f(\cdot, u_n), v >_\varrho \tag{7.1}$$

$$-G(v) \ \forall v \in S_{n.}$$

where $0 < \gamma' < \gamma$, $\gamma = (\lambda_{j0+j_1} - \lambda_{j_o})/2$, and $n \geq n_o$.

We claim there there is a constant K > 0 such that

$$\|u_n\|_{p,q,\varrho} \leq K \quad \forall n \geq n_o \tag{7.2}$$

Suppose that (7.2) is false. Then for ease of notation and without loss in generality we can assume that

$$\lim_{n\to\infty} \|u_n\|_{p,q,\varrho} = \infty \tag{7.3}$$

We shall show that (7.3) leads to a contradiction.

We first observe from (Q-3), (1.5)(iii), (1.10), and (1.11)(ii) that

$$c_2 \left[\sum_{i=1}^{N} \|D_i u_n\|_{p_i}^2 + \varepsilon_o \|u_n\|_q^2 \right] \leq \mathcal{Q}(u_n, u_n) \ \forall n \geq n_o, \tag{7.4}$$

where c_2 and ε_o are positive constants. Therefore, it follows from (7.1), (7.2), (f-3) and the fact that $G \in \left[H_{p,q,\varrho}\right]'$ that

$$\lim_{n\to\infty} \|u_n\|_{\varrho} = \infty \tag{7.5}$$

and, furthermore, that there is a constant K_1 and an $n_1 > n_o$ such that

$$\|u_n\|_{p.q,\varrho} \leq K_1 \|u_n\|_{\varrho} \quad \forall n \geq n_1. \tag{7.6}$$

Next, we write

$$u_n = u_{n1} + u_{n2} + u_{n3} \text{ and } \tilde{u}_n = -u_{n1} - u_{n2} + u_{n3} \quad \text{where}$$

$$u_{n1} = \sum_{j=1}^{j_o-1} \hat{u}_n(k)\varphi_k, \qquad u_{n2} = \sum_{k=j_o}^{j_o+j_1-1} \hat{u}_n(k)\varphi_k, \tag{7.7}$$

$$u_{n3} = \sum_{k=j_o+j_1}^{n} \hat{u}_n(k)\varphi_k.$$

$\forall n \geq max(n_1, j_o + j_1)$ and claim that

$$\lim_{n\to\infty} [\|u_{n1}\|_{p,q,\varrho} + \|u_{n3}\|_{p,q,\varrho}] / \|u_n\|_{\varrho} = 0 \tag{7.8}$$

To establish (7.8), we observe from (7.1) and (7.7) that

$$\mathcal{L}(u_n, \tilde{u}_n) - (\lambda_{j0} + \gamma') < u_n, \tilde{u}_n >_{\varrho}$$

$$= \left(1 - n^{-1}\right) < f\left(\cdot, u_n\right) - \gamma' u_n, \tilde{u}_n >_\varrho$$
$$-G(\tilde{u}_n) + \mathcal{L}\left(u_n, \tilde{u}_n\right) - \mathcal{Q}\left(u_n, \tilde{u}_n\right)$$

Hence from (7.7), we have that

$$\sum_{k=1}^{j_o+j_1-1} \left(\lambda_{j_0} + \gamma' - \lambda_k\right) |\hat{u}_n(k)|^2 + \sum_{j=j_o+j_1}^{n} \left(\lambda_k - \lambda_{j_0} - \gamma'\right) |\hat{u}_n(k)|^2 \tag{7.9}$$

$$= -G(\tilde{u}_n) + \left(1 - n^{-1}\right) < f\left(\cdot, u_n\right) - \gamma' u_n, \tilde{u}_n >_\varrho$$

$$+\mathcal{L}\left(u_n, \tilde{u}_n\right) - \mathcal{Q}\left(u_n, \tilde{u}_n\right)$$

Now we observe from (2.15) and (7.7) that

$$\left| < f\left(\cdot, u_n\right) - \gamma' u_n, \tilde{u}_n >_\varrho \right| \leq \gamma' \left\|u_n\right\|_\varrho^2 + \left\|f_o\right\|_\varrho \left\|u_n\right\|_\varrho \tag{7.10}$$

Also we set $\delta = \gamma - \gamma'$ and observe that $\delta > 0$ and that

$$\lambda_{j_0} + \gamma' = \lambda_{j_0+j_1} - (\delta + \gamma).$$

We therefore obtain from (7.9) and (7.10) that

$$\gamma' \left\|u_n\right\|_\varrho^2 + \sum_{k=1}^{j_o-1} \left(\lambda_{j_0} - \lambda_k\right) |\hat{u}_n(k)|^2 + \sum_{k=j_o+j_1}^{n} \left[\left(\lambda_k - \lambda_{j_0+j_1}\right) + \delta\right] |\hat{u}_n(k)|^2$$

$$\leq K \left\|\tilde{u}_n\right\|_{p.q,\varrho} + \gamma' \left\|u_n\right\|_\varrho^2 + \left\|f_o\right\|_\varrho \left\|u_n\right\|_\varrho + \left|\mathcal{L}\left(u_n, \tilde{u}_n\right) - \mathcal{Q}\left(u_n, \tilde{u}_n\right)\right| \tag{7.11}$$

It is clear that $\exists\ \gamma'' > 0$ such that

$$\gamma''\left(1 + \lambda_k\right) \leq \lambda_{j_0} - \lambda_k \quad \text{for } k = 1, ..., j_o - 1 \tag{7.12}$$

$$\gamma''\left(1 + \lambda_k\right) \leq \left(\lambda_k - \lambda_{j_0+j_1}\right) + \delta \quad \text{for } k \geq j_o + j_{1.}$$

Consequently, it follows from Lemma 1, (4.3), (7.11) and (7.12) that

$$\gamma^* \left[\left\|u_{n1}\right\|_{p,q,\varrho}^2 + \left\|u_{n3}\right\|_{p,q,\varrho}^2\right] \leq K_2 \left\|u_n\right\|_\varrho + K \left\|\tilde{u}_n\right\|_{p,q,\varrho} + \tag{7.13}$$

$$\left|\mathcal{L}\left(u_n, \tilde{u}_n\right) - \mathcal{Q}\left(u_n, \tilde{u}_n\right)\right|.$$

for $n \geq max(n_{1,}j_o + j_1)$ where K, K$_2$, and γ^* are positive constants.

As we have observed before $\mathcal{L}(\tilde{u}_n, \tilde{u}_n) = \mathcal{L}(u_n, u_n)$. Consequently from (4.3), (7.5), (7.6), and (1.11) (i), we have that

$$\lim_{n \to \infty} |\mathcal{L}(u_n, \tilde{u}_n) - \mathcal{Q}(u_n, \tilde{u}_n)| / \|u_n\|_\varrho^2 = 0$$

Dividing both sides of (7.13) by $\|u_n\|_\varrho^2$, using this last established fact, we see that (7.8) does indeed hold because γ^* is a positive constant.

Next, using (7.7) we set

$$u_n = w_n + v_n \qquad \text{where } w_n = u_{n2} \quad \text{and } v_n = u_{n1} + u_{n3}. \tag{7.14}$$

and observe that $< v_n, \phi_j >_\varrho = 0$ for $j = j_o, \ldots, j_o + j_1 - 1$ and also that w_n is a $\lambda_{j_o}-$ eigenfunction of L. Also from (7.8),

$$\lim_{n \to \infty} \|v_n\|_{p,q,\varrho} / \|u_n\|_\varrho = 0 \tag{7.15}$$

Setting $v = w_n$ in (7.1) and using the fact that w_n is also a $H^1_{p,q,\varrho}-$ $\lambda_{j_o}-$ eigenfunction for Q, we obtain that

$$-\gamma' n^{-1} \|w_n\|_\varrho^2 = \left(1 - n^{-1}\right) < f(\cdot, u_n), w_n >_\varrho - G(w_n) \quad \forall n \geq n_o. \tag{7.16}$$

Therefore, we infer from (7.16) that

$$\left(1 - n^{-1}\right) < f(\cdot, u_n), w_n >_\varrho - G(w_n) \ \leq 0 \ \forall n \geq n_o. \tag{7.17}$$

Consequently,

$$\lim_{n \to \infty} \sup[\left(1 - n^{-1}\right) < f(\cdot, u_n), w_n >_\varrho - G(w_n)] \leq 0$$

But from (7.5), (7.14), and (7.15), we obtain that this is a direct contradiction of (2.16). Therefore (7.2) is indeed true.

To complete the proof of the theorem, we observe from (7.2) that (5.19)-(5.24) given in the proof Theorem 1 now applies. Also, we see from (f-1), (f-3), (5.20), (5.21), and the Lebesgue dominated convergence theorem that

$$\lim_{n \to \infty} < f(\cdot, u_n), v >_\varrho = < f(\cdot, u^*), v >_\varrho \quad \forall v \in L^2_\varrho. \tag{7.18}$$

Using (5.19)-(5.24) and (7.1), it then follows in a manner similar to the proof of Theorem 1 (we leave the details to the reader) that there is a subsequence $\{u_{n_j}\}_{j=1}^\infty$ such that

$$\lim_{j \to \infty} Du_{n_j}(x) = Du^*(x) \quad \text{for } a.e.\ x \in \Omega.$$

As a consequence, we obtain (as before) that

$$\lim_{j\to\infty} \mathcal{Q}(u_{n_j}, v_j) = \mathcal{Q}(u^*, v_j) \text{ for } v_J \in S_J \text{ and } J \geq n_o.$$

But this fact joined with (7.1) and (7.18) enables us to obtain that

$$\mathcal{Q}(u^*, v_J) = \lambda_{j_o} < u^*, v_J >_\varrho + < f(\cdot, u^*), v_J >_\varrho \quad (7.19)$$

$$-G(v_J) \quad \forall v_J \in S_J, J \geq n_o.$$

Next, given $v \in H^1_{p,q,\varrho}$, we define $P_J v$ as in (5.37) and observe

$$\mathcal{Q}(u^*, P_J v) \to \mathcal{Q}(u^*, v), \quad < u^*, P_J v >_\varrho \to < u^*, v >_\varrho,$$

$$< f(\cdot, u^*), P_J v >_\varrho \to \; < f(\cdot, u^*), v >_\varrho, \text{ and } G(v_J) \to G(v).$$

Hence it follows from these facts that (7.19) holds with v replacing v_J, and the proof of Theorem 5 is complete.

1.8 Proofs of Theorem 6 and Corollary 7

We prove Theorem 6 first. Without loss of generality we can assume from the start that $j_o > 1$. We set $\gamma' = \gamma/2$, and observe that since f meets (f-5) it also meets (f-3) with the γ' just selected as well as (f-4). Hence the conditions in the hypothesis of Theorem 6 imply the conditions in the hypothesis Lemma 5. Consequently we can invoke Lemma 5 (as we did in the proof Theorem 5) and obtain a sequence of functions $\{u_n\}_{n=n_o}^\infty \subset H^1_{p,q,\varrho}$ such that

$$\mathcal{Q}(u_n, v) = (\lambda_{j0} + \gamma' n^{-1}) < u_n, v >_\varrho + (1 - n^{-1}) < f(\cdot, u_n), v_J >_\varrho - G(v) \quad (8.1)$$

$\forall v \in S_n$ where $\gamma' = \gamma/2, \gamma = (\lambda_{j0+j_1} - \lambda_{j_o})/2$ and n$\geq n_o$.
We claim there is a constant $K > 0$ such that

$$\|u_n\|_{p,q,\varrho} \leq K \quad \forall n \geq n_o. \quad (8.2)$$

Suppose that (8.2) does not hold. Then for ease of notation and without loss in generality, we can assume that

$$\lim_{n\to\infty} \|u_n\|_{p,q,\varrho} = \infty. \quad (8.3)$$

Then in a manner similar to that in the proof of Theorem 5, we arrive at the fact that

$$\lim_{n\to\infty} \|u_n\|_\varrho = \infty \quad (8.4)$$

and that there is an $n_1 \geq n_o$ and a constant K_1 such that

$$\|u_n\|_{p,q,\varrho} \leq K_1 \|u_n\|_{\varrho} \quad \text{for } n \geq n_1. \tag{8.5}$$

Next, we write

$$u_n = u_{n1} + u_{n2} + u_{n3} \quad \forall n \geq max(n_1, j_o + j_1) \text{ where} \tag{8.6}$$

$$u_{n1} = \sum_{j=1}^{j_o-1} \hat{u}_n(k)\varphi_k, \qquad u_{n2} = \sum_{k=j_o}^{j_o+j_1-1} \hat{u}_n(k)\varphi_k,$$

$$u_{n3} = \sum_{k=j_o+j_1}^{n} \hat{u}_n(k)\varphi_k.$$

and obtain exactly as in the proof Theorem 5 that

$$\lim_{n\to\infty} [\|u_{n1}\|_{p,q,\varrho} + \|u_{n3}\|_{p,q,\varrho}] / \|u_n\|_{\varrho} = 0. \tag{8.7}$$

Now, from (8.6), it follows that u_{n2} is a λ_{j_o}-eigenfunction of L and furthermore that $P^o_{j_o} u_n = u_{n2}$. Also, we have from the hypothesis of Theorem 6 that λ_{j_o} is an L-pseudo-eigenvalue of Q. Consequently, we infer from (8.3), (8.5), (8.7), and (2.18) that

$$\lim_{n\to\infty} [\mathcal{Q}(u, u_{n2}) - \mathcal{L}(u_n, u_{n2})] / \|u_n\|_{\varrho} = 0. \tag{8.8}$$

Next, we obtain from (8.1) that

$$-\gamma' n^{-1} \|u_{n2}\|^2_{\varrho} = (1 - n^{-1}) < f(\cdot, u_n), v_J >_{\varrho} -G(u_{n2})$$
$$+\mathcal{L}(u_n, u_{n2}) - Q(u_n, u_{n2}).$$

We infer from this last inequalilty that

$$(1 - n^{-1}) < f(\cdot, u_n), u_{n2} >_{\varrho} \leq G(u_{n2}) + Q(u_n, u_{n2}) - \mathcal{L}(u_n, u_{n2}) \tag{8.9}$$

and set

$$W_n = u_{n2} / \|u_n\|_{\varrho} = \sum_{k=j_o}^{j_o+j_1-1} \hat{u}_n(k)\varphi_k / \|u_n\|_{\varrho}. \tag{8.10}$$

As a consequence of the above, in particular (8.7), we see that there is a subsequence (which for ease of notation we take to be the full sequence) and a λ_{j_o}-eigenfunction W of L such that

$$(i)\ W = \sum_{k=j_o}^{j_o+j_1-1} \alpha_k\varphi_k \text{ where } \sum_{k=j_o}^{j_o+j_1-1} \alpha_k^2 = 1, \tag{8.11}$$

$$(ii)\ \lim_{n\to\infty} \mathrm{W}_n(x) = W(x) \quad \text{a.e. in } \Omega,$$

$$(iii)\ \lim_{n\to\infty} \left\| u_n/\left\|u_n\right\|_\varrho - W \right\|_\varrho = 0,$$

$$(iv)\ \lim_{n\to\infty} u_n(x) / \left\|u_n\right\|_\varrho = W(x) \quad \text{a.e. in } \Omega,$$

$$(v)\ \lim_{n\to\infty} \left\|W_n - W\right\|_{p,q,\varrho} = 0,$$

$$(vi)\ \exists W^* \in L\varrho^2 \text{ s.t. } |W_n(x)| \le W^*(x) \quad \text{a.e. in } \Omega.$$

Also from (f-5) and (8.11)(vi), we see that there is a constant K_2 such that

$$\int_\Omega |f(x,u_n)\, W_n(x)|\, \varrho \le K_2 \quad \forall n. \tag{8.12}$$

Dividing both sides of (8.9) by $\left\|u_n\right\|_\varrho$ and leaving $n \to \infty$, it follows from (8.8), (8.10), (8.11)(v), and (8.12) that

$$\lim_{n\to\infty} \sup < f(\cdot,u_n), W_n >_\varrho \le G(W). \tag{8.13}$$

Next, we set

$$A = \{x \in \Omega : W(x) > 0\} \text{ and } B = \{x \in \Omega : W(x) < 0\} \tag{8.14}$$

Then it follows from (8.11))(ii)and (vi), (f-5), and (8.13) that

$$\lim_{n\to\infty} \inf \int_A f(x,u_n) W_n \varrho + \lim_{n\to\infty} \inf \int_B f(x,u_n) W_n \varrho \le G(W). \tag{8.15}$$

Now, it follows from (2.19), (8.4), and (8.11)(ii) and (iv)that

$$(i)\quad \lim_{n\to\infty} \inf f(x,u_n)\, W_n(x) \ge f_+(x)\, W(x) \quad \text{for a.e. } \mathrm{x} \in A, \tag{8.16}$$

$$(ii)\quad \lim_{n\to\infty} \inf f(x,u_n)\, W_n(x) \ge f^-(x)\, W(x) \quad \text{for a.e. } \mathrm{x} \in B.$$

Also, we have from (f-5) and (8.11)(vi) that for $n \ge n_1$,

$$-f_o(x) \quad W^*(x) \le f(x,u_n)\, W_n(x) \qquad \text{for a.e.} \mathrm{x} \in \Omega,$$

where f_o and W^* are in $\mathrm{L}\varrho^2$. We consequently conclude from (8.15), (8.16), and Fatou's Lemma [Rud, p. 24], that

$$\int_A f_+(x)\, W(x)\, \varrho(x)\, dx + \int_B f^-(x)\, W(x)\, \varrho(x)\, dx \le G(W). \tag{8.17}$$

But by (8.11)(i), W is a nontrivial $\lambda_{j_o}-$eigenfunction of L. Consequently, we see from (8.14) that (8.17) is a contradiction to (2.20). We conclude that (8.3) is false and that (8.2) is indeed true.

The conclusion of the proof of Theorem 6 follows along the same lines as the proof of Theorem 5. We leave the details to the reader.

The proof of the sufficiency condition in Corollary 7 follows immediately from Theorem 6. It remains to establish the necessary condition of Corollary 7. Hence assume that $u^* \in H^1_{p,q,\varrho}$ is a weak solution of (2.17) and that W is a nontrivial λ_{j_o}-eigenfunction of L. Consequently, it then follows that

$$\mathcal{Q}\,(u^*, W)) = \lambda_{j0} < u^*, W >_\varrho + < f(\cdot, u^*), W >_\varrho - G(W). \qquad (8.18)$$

Since by assumption every $\lambda_{j_o}-$ eigenfunction of L is a $\lambda_{j_o}-$ eigenfunction of Q, we then obtain from (8.18) that

$$\int_A f\,(x, u^*)\, W \varrho + \int_B f\,(x, u^*)\, W \varrho = G(W). \qquad (8.19)$$

where A and B are defined in (8.14). From (2.21), we see

$$\int_A f\,(x, u^*)\, W \varrho < \int_A f_+ W \varrho \quad \text{if} \int_A W \varrho > 0. \qquad (8.20)$$

Likewise, we see from (2.21) that

$$\int_B f\,(x, u^*)\, W \varrho < \int_B f^- W \varrho \quad \text{if} \int_B W \varrho < 0. \qquad (8.21)$$

Also, we see from (f-5) and (8.14)that

$$\int_A f_+ W \varrho = \int_A f\,(x, u^*)\, W \varrho \ = 0 \ \text{ if} \int_A W \varrho = 0.$$

and that

$$\int_B f^- W \varrho = \int_B f\,(x, u^*)\, W \varrho \ = 0 \ \text{ if} \int_B W \varrho = 0$$

Now by assumption W is a nontrivial $\lambda_{j_o}-$eigenfunction of L; so either $\int_A W\varrho > 0$ or $\int_B W\varrho < 0$ or both hold. Hence it follows from (8.20) and (8.21) that

$$\int_A f\,(x, u^*)\, W \varrho + \int_B f\,(x, u^*)\, W \varrho < \int_A f_+ W \varrho + \int_B f^- W \varrho$$

This last inequality in conjunction with (8.19) gives (2.20) and establishes the necessary condition of Corollary 7.

1.9 Proof of Theorem 8

Without loss in generality, we assume from the start that $j_o > 1$. The conditions in the hypothesis of Theorem 8 imply those in the hypothesis of Lemma 6. So we invoke this lemma and obtain a sequence of functions $\{u_n\}_{n=n_1}^{\infty}$ with $u_n \in S_n$ where

$$n_1 \geq j_o + j_1 \quad \text{and} \quad \mathrm{n}_1 \geq 2/[\lambda_{j_o} - \lambda_{j_o-1}] \tag{9.1}$$

such that

$$\mathcal{Q}(u_n, v) = (\lambda_{j_o} - n^{-1}) < u_n, v > \varrho + (1+n^{-1}) < f(\cdot, u_n), v >_\varrho - G(v) \tag{9.2}$$

$$\forall v \in S_n.$$

We claim there is a constant K_1 such that

$$\|u_n\|_{p,q,\varrho} \leq K_1 \quad \text{for } n \geq n_1. \tag{9.3}$$

Suppose that (7.3) does not hold. Then without loss in generality, we can assume

$$\lim_{n \to \infty} \|u_n\|_{p,q,\varrho} = \infty. \tag{9.4}$$

Taking $v = u_n$ in (9.2), we conclude in a manner similar to that used in the proof of Theorem 5 that

$$\lim_{n \to \infty} \|u_n\|_\varrho = \infty, \tag{9.5}$$

and using the fact that $G \in H^1_{p,q,\varrho}$, we obtain, furthermore, that

$$\|u_n\|_{p,q,\varrho} \leq K_2 \|u_n\|_\varrho \quad \forall n \geq n_1 \tag{9.6}$$

where K_2 is a constant.

To show that (9.4) leads to a contradiction, we write u_n and $\tilde{u}$ as in (7.7) and obtain from (9.2) that

$$\begin{aligned}
&\mathcal{L}(u_n, \tilde{u}_n) - \lambda_{j0} < u_n, \tilde{u}_n >_\varrho \\
&= -n^{-1} < u_n, \tilde{u}_n >_\varrho + \left(1 + n^{-1}\right) < f(\cdot, u_n),, \tilde{u}_n >_\varrho \\
&-G(\tilde{u}_n) + \mathcal{L}(u_n, \tilde{u}_n) - \mathcal{Q}(u_n, \tilde{u}_n).
\end{aligned}$$

Hence, since λ_{j_o} is an eigenfunction of L of multiplicity j_1,

$$\sum_{k=1}^{j_o-1} (\lambda_{j_o} - \lambda_k) |\hat{u}_n(k)|^2 + \sum_{k=j_o+j_1}^{n} (\lambda_k - \lambda_{j_o}) \ |\hat{u}_n(k)|^2 \tag{9.7}$$

$$\leq n^{-1} \|u_n\|_\varrho^2 + (1+n^{-1}) \| f(\cdot, u_n) \|_\varrho \|u_n\|_\varrho + | G(\tilde{u}_n) |$$
$$+ | \mathcal{Q}(u_n, \tilde{u}_n) - \mathcal{L}(u_n, \tilde{u}_n) |$$

Using (5.12)(i) and (ii) and (4.3) we see that the left-hand side of (9.7) majorizes

$$\gamma_3[\|u_{n1}\|_{p,q,\varrho} + \|u_{n3}\|_{p,q,\varrho}]$$

where $\gamma_3 > 0$. So we conclude from (f-7), the fact that $G \in [H^1_{p,q,\varrho}]'$, and (9.6), on dividing both sides of (9.7) by $\|u_n\|_\varrho^2$ and passing to the limit as n$\to \infty$, that

$$\gamma_3 \lim_{n\to\infty} \sup[\|u_{n1}\|_{p,q,\varrho} + \|u_{n3}\|_{p,q,\varrho}]^2 / \|u_n\|_\varrho^2$$

$$\leq \lim_{n\to\infty} \sup |\mathcal{Q}(u_n, \tilde{u}_n) - \mathcal{L}(u_n, \tilde{u}_n)| / \|u_n\|_\varrho^2$$

But Q is #-#-related to L. So it follows from (2.22), (4.3), and (9.6) that the right-side of this last inequality is zero. As a consequence, we have, since $\gamma_3 > 0$, that

$$\lim_{n\to\infty} [\|u_{n1}\|_{p,q,\varrho} + \|u_{n3}\|_{p,q,\varrho}] / \|u_n\|_\varrho = 0. \tag{9.8}$$

Now from (7.7), $u_n = u_{n1} + u_{n2} + u_{n3}$ where u_{n2} is a λ_{j_o} - eigenfunction of L. Hence, using the notation of (2.18), we see that $P^o_{j_o} u_n = u_{n2}$. So from (9.8), we have that $\left\| u_n - P^o_{j_o} u_n \right\|_{p.q,\varrho} / \|u_n\|_{p,q,\varrho} \to 0$. However, λ_{j_o} is an L–pseudo-eigenvalue of Q, and we therefore obtain from (2.18) and (9.6) that

$$\lim_{n\to\infty} |\mathcal{Q}(u_n, u_{n2}) - \mathcal{L}(u_n, u_{n2})| / \|u_n\|_\varrho = 0. \tag{9.9}$$

Next, we observe that

$$\mathcal{Q}(u_n, u_n) - \mathcal{L}(u_n, u_n) = \mathcal{Q}(u_n, u_{n1} + u_{n3}) - \mathcal{L}(u_n, u_{n1} + u_{n3}) \tag{9.10}$$
$$+ \mathcal{Q}(u_n, u_{n2}) - \mathcal{L}(u_n, u_{n2})$$

But by the #-#-relationship of Q to L, we see that

$$|\mathcal{Q}(u_n, u_{n1} + u_3) - \mathcal{L}(u_n, u_{n1} + u_3)|$$

$$\leq K[\|u_{n1}\|_{p,q,\varrho} + \|u_{n3}\|_{p,q,\varrho} \qquad \forall n \geq n_1.$$

So we conclude from (9.8), (9.9), (9.10), and this last inequality that

$$\lim_{n\to\infty} |\mathcal{Q}(u_n, u_n) - \mathcal{L}(u_n, u_n)| / \|u_n\|_\varrho = 0. \tag{9.11}$$

Next, we put u_n in place of v in (9.2) and obtain

$$\sum_{k=1}^{n} \lambda_k \left|\hat{u}_n(k)\right|^2 - \left(\lambda_{j_o} - n^{-1}\right) \sum_{k=1}^{n} \left|\hat{u}_n(k)\right|^2$$

$$= \left(1+n^{-1}\right) < f(\cdot, u_n), u_n >_\varrho - G(u_n) + \mathcal{L}(u_n, u_n) - Q(u_n, u_n).$$

As a consequence, for $n \geq n_1$ (see (9.1)),

$$\sum_{k=1}^{j_o-1} [\lambda_{j_o} - \lambda_k - n^{-1}]^2 \left|\hat{u}_n(k)\right|^2 \geq \tag{9.12}$$

$$-\left(\lambda_{j_o} - \lambda_{j_o-1} - n^{-1}\right)[\left(1+n^{-1}\right) < f(\cdot, u_n), u_n >_\varrho - G(u_n)]$$

$$-(\lambda_{j_o} - \lambda_{j_o-1} - n^{-1})[\mathcal{L}(u_n, u_n) - Q(u_n, u_n)]$$

From (9.2), we also obtain

$$(\lambda_k + n^{-1} - \lambda_{j_o})\hat{u}_n(k) = \left(1+n^{-1}\right) < f(\cdot, u_n), \varphi_k >_\varrho - G(\varphi_k)$$

$$+\mathcal{L}(u_n, \varphi_k) - Q(u_n, \varphi_k).$$

Consequently, upon squaring we obtain

$$\begin{aligned} &(\lambda_k + n^{-1} - \lambda_{j_o})^2 \left|\hat{u}_n(k)\right|^2 \\ &\leq \left(1+n^{-1}\right)^2 < f(\cdot, u_n), \varphi_k >_\varrho^2 + \mid G(\varphi_k) \mid^2 \\ &+ \left|\mathcal{L}(u_n, \varphi_k) - Q(u_n, \varphi_k)\right|^2 + 2\left(1+n^{-1}\right) \left|< f(\cdot, u_n), \varphi_k >_\varrho\right| \left|G(\varphi_k)\right| \\ &+2\left(1+n^{-1}\right) \left|< f(\cdot, u_n), \varphi_k >_\varrho\right| \left|\mathcal{L}(u_n, \varphi_k) - Q(u_n, \varphi_k)\right| \\ &+2\left|G(\varphi_k)\right| \left|\mathcal{L}(u_n, \varphi_k) - Q(u_n, \varphi_k)\right|. \end{aligned} \tag{9.13}$$

Observing that $2|a|\,|b| \leq \varepsilon^2 a^2 + b^2/\varepsilon^2$ for a,b $\in \mathbf{R}$ and that there is a constant K_3 such that $|G(\varphi_k)| \leq K_3$ for $k = 1, \ldots, j_o - 1$, we obtain from (9.12) and (9.13) that

$$-(\lambda_{j_o} - \lambda_{j_o-1} - n^{-1})[\left(1+n^{-1}\right) < f(\cdot, u_n), u_n >_\varrho - G(u_n)$$

$$+\mathcal{L}(u_n, u_n) - Q(u_n, u_n)]$$

$$\leq \left(1+n^{-1}\right)\left(1+2\varepsilon^2\right) \|f(\cdot, u_n)\|_\varrho^2 + (1 + \frac{2}{\varepsilon^2})K_3^2 j_o \tag{9.14}$$

$$+\left(1+\varepsilon^{2}+\varepsilon^{-2}\right)\sum_{k=1}^{j_o-1}\left|\mathcal{L}\left(u_n,\varphi_k\right)-Q\left(u_n,\varphi_k\right)\right|^2$$

for $n\geq n_1$ and $\varepsilon>0$.

Now from (f-6), we have that $-sf\left(x,s\right)\geq -f_o\left(x\right)\left|s\right|$. Therefore

$$-sf\left(x,s\right)\geq\left|s\right|\left|\, f\left(x,s\right)\right|-2f_o\left(x\right)\left|s\right|.$$

Also from (f-7), we have that for $\delta>0$ and $s\in\mathbf{R}$,

$$\delta\left|s\right|\geq\left|\, f\left(x,s\right)\right|-b_\delta\left(x\right).$$

where $b_\delta\in L^2_\varrho\left(\Omega\right)$. We consequently obtain from this last inequality that

$$-\delta s\ f\left(x,s\right)\geq\left(1-\delta\right)\left|f\left(x,s\right)\right|^2-b_\delta^2\left(x\right)/4\delta-2\delta f_o\left(x\right)\left|s\right|$$

and therefore that

$$\left|f\left(x,s\right)\right|^2\leq-\delta s\ f\left(x,s\right)/\left(1-\delta\right) \tag{9.15}$$
$$+b_\delta^2\left(x\right)/4\delta\left(1-\delta\right)+2\delta f_o\left(x\right)\left|s\right|/\left(1-\delta\right)$$

We use (9.15) in conjunction with (9.14) where $\varepsilon=1$ and δ is a fixed but arbitrary positive number to obtain

$$\begin{aligned}&\left(\lambda_{j_o}-\lambda_{j_o-1}-n^{-1}\right)G(u_n)\leq\\ &\quad\left(1+n^{-1}\right)\left[\left(\lambda_{j_o}-\lambda_{j_o-1}-n^{-1}\right)-3\delta/\left(1-\delta\right)\right]<f\left(\cdot,u_n\right),u_n>_\varrho\\ &\quad+\left(1+n^{-1}\right)3\left[\left\|b_\delta\left(x\right)\right\|_\varrho^2/4\delta\left(1-\delta\right)+2\delta<f_o,\left|\,u_n\,\right|>_\varrho/\left(1-\delta\right)\right]\\ &+3\sum_{k=1}^{j_o-1}\left|\mathcal{L}\left(u_n,\varphi_k\right)-Q\left(u_n,\varphi_k\right)\right|\ +3K_3^2j_o\\ &\quad+\left(\lambda_{j_o}-\lambda_{j_o-1}-n^{-1}\right)\left[\mathcal{L}\left(u_n,u_n\right)-Q\left(u_n,u_n\right)\right].\end{aligned} \tag{9.16}$$

Next, we set $W_n=u_n/\left\|u_n\right\|_\varrho$ and observe from (7.7) and (9.8) that there exists a $\lambda_{j_o}-$eigenfunction W and a subsequence (which for ease of notation we take to be the full sequence) such that

$$(i)\ \lim_{n\to\infty}\left\|W_n-W\right\|_{p,q,\varrho}=0 \tag{9.17}$$

$$(ii)\quad \lim_{n\to\infty}\mathrm{W}_n(x)=W\left(x\right)\quad\text{a.e. in }\Omega.$$

We divide both sides of (9.16) by $\|u_n\|_\varrho$ and leave $n \to \infty$ to obtain from (9.11), (9.17), and the fact that Q is #-#-related to L that

$$(\lambda_{j_o} - \lambda_{j_o-1})\, G(W) \leq$$

$$[(\lambda_{j_o} - \lambda_{j_o-1}) - 3\delta/\,(1-\delta)] \lim_{n\to\infty} \inf < f\,(\cdot, u_n)\,, W_n >_\varrho$$

$$+6\delta < f_o, |\,W\,| >_\varrho /\,(1-\delta)\,.$$

We then leave $\delta \to 0$ in this last inequality and obtain

$$G(W) \leq \lim_{n\to\infty} \inf < f\,(\cdot, u_n)\,, W_n >_\varrho \cdot \tag{9.18}$$

Since

$$\|W_n\|_\varrho = \|u_n\|_\varrho /\, \|u_n\|_\varrho = 1,$$

we have from (9.17)(i) that $\|W\|_\varrho = 1$. Therefore,

$$W \text{ is a nontrivial } \lambda_{j_o} - \text{eigenfunction of } L.$$

Also, we have from (9.17)(i), that there is $W^*(x) \in L^2_\varrho$ such that

$$|W_n\,(x)| \leq W^*(x) \quad \text{a.e. in } \Omega.$$

Hence in particular we have (f-6) that

$$f(x, u_n)W_n(x) \leq f_o\,(x)\,|W_n(x)| \leq f_o\,(x)\,W^*(x) \quad \text{a.e.in } \Omega. \tag{9.19}$$

We set

$$A = \{x \in \Omega : W(x) > 0\},$$

$$B = \{x \in \Omega : W(x) < 0\}$$

$$C = \{x \in \Omega : W(x) = 0\}$$

Consequently, we have from (9.19) and Fatou's lemma that the right-side of (9.18) is majorized by

$$\{\int_A + \int_B + \int_C\} \left[\lim_{n\to\infty} \sup\, f(x, u_n)W_n\right] \varrho dx.$$

Now an easy computation shows that

$$\lim_{n\to\infty} \sup f(x,u_n)W_n(x) \le f^+(x)\,W(x) \quad \text{a.e. in } A,$$

$$\lim_{n\to\infty} \sup f(x,u_n)W_n(x) \le f_-(x)\,W(x) \quad \text{a.e.in } B,$$

$$\lim_{n\to\infty} \sup f(x,u_n)W_n(x) \le 0 \quad \text{a.e. in } C.$$

We conclude from (9.18) that

$$G(W) \le \int_A f^+ W\varrho + \int_B f_- W\varrho.$$

Since W is a nontrivial $\lambda_{j_o}-$eigenfunction of L, we see that this last inequality is a direct contradiction of (2.23). Hence (9.4) does not hold and (9.3) is indeed true.

The rest of the proof of Theorem 8 proceeds along the same lines as Theorems 1 and 5, and we leave the details to the reader.

1.10 Proof of Theorem 9

To prove Theorem 9, we suppose we are given a sequence of functions $\{v_n\}_{n=1}^{\infty}$ in $H^1_{p,q,\varrho}(\Omega)$, where (Ω,Γ) is a Simple V_L-region and

$$\|v_n\|_{p,q,\varrho} \le K \qquad \forall n \tag{10.1}$$

where K is a finite constant. What we have to show is that given θ with $2<\theta<2N/(N-1)$, $\exists\ v \in H^1_{p,q,\varrho}$ and a subsequence $\{v_{n_j}\}_{j=1}^{\infty}$ such that

$$\lim_{j\to\infty} \int_\Omega \left|v_{n_j} - v\right|^\theta \varrho dx = 0 \tag{10.2}$$

It is clear from the definition of $H^1_{p,q,\varrho}$, we can assume from the start that

$$v_n \in C^1_{p,q,\varrho}(\Omega) \qquad \forall n. \tag{10.3}$$

Also, it follows from Lemma 7 that $v_n, v \in H^1_{p,q,\varrho}$ implies that $v_n, v \in L^\theta_\varrho$. Hence, (10.2) is a feasible condition.

To establish (10.2), we observe from Lemma 2 that $H^1_{p,q,\varrho}$ is compactly imbedded in L^2_ϱ. Consequently, it follows from the proof of Lemma 2 that there is a $v \in H^1_{p,q,\varrho} \cap L^2_\varrho \cap L^\theta_\varrho$ and a subsequence (which we take to be the full sequence) such that $\|v_n - v\|_\varrho \to 0$ and furthermore such that

$$\lim_{n\to\infty} v_n(x) = v(x) \quad \text{for a.e. } x\in\Omega. \tag{10.4}$$

Now if we can show that $\{| v_n(x) |^\theta\}_{n=1}^\infty$ is absolutely equiintegrable with respect to the measure μ_ϱ where $\mu_\varrho(E) = \int_E \varrho dx$, i.e., given $\varepsilon > 0$, $\exists \delta > 0$ such that $\mu_\varrho(E) < \delta$ implies that

$$\int_E | v_n(x) |^\theta \varrho dx < \varepsilon \quad \forall n, \tag{10.5}$$

then (10.2) will follow from (1.1), (10.4), and Egoroff's theorem [Rud, p.75]. (For a proof of this fact, see [DS, p.325].)

So to complete the proof the theorem, all we have to show is that (10.5) holds. Without loss in generality we assume that N$\geq$ 2 and proceed to establish (10.5). It will be clear that a similar situation prevails for N=1.

By assumption, $\Omega = \Omega_1 \times \cdots \times \Omega_N$ and we see from (10.3) for fixed $x_1, \cdots, x_{i-1}, x_{i+1}, \cdots, x_N$, $v_n(x)$ as a function of x_i is in $C^1(\Omega_i)$. Therefore by (2.25) (iv), $|v_n(x)| \leq h_i(x_i)\, g_i^{\frac{1}{2}}(x_1, \cdots, x_{i-1}, x_{i+1}, \cdots, x_N)$ where

$$g_i(x_1, \cdots, x_{i-1}, x_{i+1}, \cdots, x_N) = \int_{\Omega_i} [p_i^*(x_i) |D_i v_n(x)|^2 + \varrho_i^*(x_i) |v_n(x)|^2] dx_i \tag{10.6}$$

and where $h_i(x_i) \in L^\gamma_{\varrho_i^*}(\Omega_i)$ for every γ with $2 < \gamma < \infty$. As a consequence,

$$|v_n(x)|^\theta \leq \left[\prod_{i=1}^N h_i(x_i)\right]^{\theta/N} \left[\prod_{i=1}^N g_i(\check{x}_i)\right]^{\theta/2N} \quad \forall x \in \Omega. \tag{10.7}$$

where $\check{x}_i = (x_1, \cdots, x_{i-1}, x_{i+1}, \cdots, x_N)$.

Since $\varrho(x) = \varrho_1^*(x_i) \cdots \varrho_N^*(x_N)$, we observe that

$$h(x) = \prod_{i=1}^N h_i(x_i) \text{ is s.t. } h \in L^\gamma_\varrho(\Omega) \text{ for } 2 < \gamma < \infty. \tag{10.8}$$

Taking $r_2 = 2N/(N-1)\theta$ and $r_1 = r_2/(r_2 - 1)$, we therefore have from (10.7) and Holder's inequality that for $E \subset \Omega$, a Lebesgue measurable set, that

$$\int_E |v_n|^\theta \varrho(x)\, dx \leq \left\{\int_E |h(x)|^{\theta r_1/N} \varrho(x)\, dx\right\}^{1/r_1} \tag{10.9}$$

$$\left\{\int_\Omega \left[\prod_{i=1}^N g_i(\check{x}_i)\right]^{1/(N-1)} \varrho(x)\, dx\right\}^{1/r_2}$$

Now from (10.8), we see that $|h|^{\theta r_1/N} \in L^1_\varrho(\Omega)$. Consequently, we can choose $\delta > 0$ such that if $\mu_\varrho(E) < \delta$, then

$$\left\{\int_E |h|^{\theta r_1/N} \varrho(x)\,dx\right\}^{1/r_1} < \varepsilon/K^\theta \tag{10.10}$$

where K is given in (10.1). We claim

$$\left\{\int_\Omega \left[\prod_{i=1}^N g_i(\check{x}_i)\right]^{1/(N-1)} \varrho(x)\,dx\right\}^{1/r_2} \le K^\theta. \tag{10.11}$$

Hence it follows from (10.9), (10.10), and (10.11) that if $\mu_\varrho(E) < \delta$, then $\int_E |v_n|^\theta \varrho(x)\,dx < \varepsilon \quad \forall n$, which is (10.5) the desired result. So the proof of the theorem will be complete once (10.11) is established.

To establish (10.11), we first observe from (10.6) and the generalized Holder's inequality, [Zy, p. 18], that

$$\int_{\Omega_N} [g_1 ... g_{N-1} g_N]^{1/(N-1)} \varrho^*_N(x_N)\,dx_N$$

$$\le g_N^{1/(N-1)}(\check{x}_N) \prod_{i=1}^{N-1} \left[\int_{\Omega_N} g_i(\check{x}_i)\,\varrho^*_N(x_N)\,dx_N\right]^{1/(N-1)}$$

and that

$$\int_{\Omega_{N-1}} \int_{\Omega_N} [g_1 ... g_{N-1} g_N]^{1/(N-1)} \varrho^*_{N-1}(x_{N-1})\,\varrho^*_N(x_N)\,dx_N dx_{N-1} \le$$

$$\left[\int_{\Omega_N} g_{N-1}(\check{x}_{N-1})\,\varrho^*_N(x_N)\,dx_N\right]^{1/(N-1)}$$

$$\left[\int_{\Omega_{N-1}} g_N(\check{x}_N)\,\varrho^*_{N-1}(x_{N-1})\,dx_{N-1}\right]^{1/(N-1)}$$

$$\prod_{i=1}^{N-2} \left[\int_{\Omega_{N-1}} \int_{\Omega_N} g_i(\check{x}_i)\,\varrho^*_{N-1}(x_{N-1})\,\varrho^*_N(x_N)\,dx_{N-1} dx_N\right]^{1/(N-1)}$$

We set $\check{\Omega}_i = \Omega_1 \times \cdots \times \Omega_{i-1} \times \Omega_{i+1} \times \cdots \times \Omega_N$ and continue the above iterative procedure to obtain

$$\int_{\Omega_1} \cdots \int_{\Omega_{N-1}} \int_{\Omega_N} [g_i(\check{x}_i) \cdots g_{N-i}(\check{x}_i)\, g_N(\check{x}_N)]^{1/(N-1)} \tag{10.12}$$

$$\varrho_1^*(x_1)\cdots\varrho_{N-1}^*(x_{N-1})\,\varrho_N^*(x_N)\,dx_1\cdots dx_{N-1}dx_N \le$$

$$\prod_{i=1}^{N}[\int_{\check{\Omega}_i} g_i(\check{x}_i)\,\varrho_1^*(x_1)\cdots\varrho_{i-1}^*(x_{i-1})\,\varrho_{i+1}^*(x_{i+1})$$

$$dx_1\cdots dx_{i-1}dx_{i+1}\cdots dx_N]^{1/(N-1)}.$$

Now from (10.6) and the fact that

$$p_i(x) = \varrho_1^*(x_1)\cdots\varrho_{i-1}^*(x_{i-1})\,p_i^*(x_i)\varrho_{i+1}^*(x_{i+1})\cdots\varrho_N^*(x_N),$$

and that $\varrho(x) = \varrho_1^*(x_1)\cdots\varrho_N^*(x_N)$, we see that

$$\int_{\check{\Omega}_i} g_i(\check{x}_i)\,\varrho_1^*(x_1)\cdots\varrho_{i-1}^*(x_{i-1})\,\varrho_{i+1}^*(x_{i+1})\,dx_1\cdots dx_{i-1}dx_{i+1}\cdots dx_N.$$

$$= \int_{\Omega}\left[p_i(x)\,|D_i v_n(x)|^2 + \varrho(x)\,|v_n(x)|^2\right]dx \le \|v_n\|_{p,\varrho}^2 \le \|v_n\|_{p,q,\varrho}^2.$$

Therefore we obtain from (10.1), and (10.12) that

$$\int_{\Omega}\left[\prod_{i=1}^{N} g_i(\check{x}_i)\right]^{1/(N-1)}\varrho(x)\,dx_1\cdots dx_N \le \left[\|v_n\|_{p,q,\varrho}^{2/(N-1)}\right]^N \le K^{2N/(N-1)}$$

But it then follows from this last inequality that the left-hand side of (10.11) is majorized by $K^{2N/r_2(N-1)}$. Now checking back we see that $r_2 = 2N/(N-1)\theta$. Therefore, the left-hand side of (10.11) is majorized by K^θ, and the proof of the theorem is complete.

1.11 Proofs of Theorems 10, 11, and 12

To prove Theorem 10, we invoke Lemma 8 and obtain a sequence $\{u_n\}_{n=2}^\infty$ where

$$u_n = \alpha_1^n\varphi_1 + \cdots + \alpha_n^n\varphi_n \qquad (11.1)$$

with $\alpha_j^n \in \mathbf{R}$, $j = 1,\cdots,n$ and $\{\varphi_n\}_{n=1}^\infty$ is the sequence given in (V_L -1) and (V_L -2) with the property that

$$\mathcal{Q}(u_n, v) = (\lambda_1 - n^{-1}) < u_n, v >_\varrho + \int_\Omega f(x, u_n)v\varrho - G(v) \quad \forall v \in S_n. \quad (11.2)$$

We claim there is a constant K such that

$$\|u_n\|_{p,q,\varrho} \leq K \qquad \forall n. \tag{11.3}$$

Suppose to the contrary that (11.3) does not hold. Then without loss in generality, we assume

$$\lim_{n\to\infty} \|u_n\|_{p,q,\varrho} = \infty. \tag{11.4}$$

We will show that this assumption gives rise to a contradiction.

To accomplish this contradiction, putting u_n for v in (11.2), we see that

$$\mathcal{Q}(u_n, u_n) = (\lambda_1 - n^{-1}) < u_n, u_n >_\varrho + \int_\Omega f(x, u_n)u_n\varrho - G(u_n). \tag{11.5}$$

Next, we observe from (11.1) and Lemma 1 (with $\hat{u}_n(j) = \alpha_j^n$) that

$$\mathcal{L}(u_n, u_n) = \sum_{j=1}^{n} |\hat{u}_n(j)|^2 \lambda_j \geq \lambda_1 \|u_n\|_\varrho^2 .$$

Consequently, we obtain from (11.5), (2.28), and Holder's inequality that

$$\begin{aligned} \varepsilon_o^* \|u_n\|_\varrho^2 &\leq \{\int_\Omega \left|\tilde{h}_o\right|^{\theta^*} \varrho\}^{1/\theta^*} \{\int_\Omega |u_n|^\theta \varrho\}^{1/\theta} \\ &+ |G(u_n)| + |\mathcal{Q}(u_n, u_n) - \mathcal{L}(u_n, u_n)| \quad \forall n. \end{aligned}$$

But then it follows Lemma 7, the fact that $G \in [H^1_{p,q,\varrho}]'$, the #-relationship between Q and L, ε_o^* being strictly positive, and this last inequality that

$$\lim_{n\to\infty} \|u_n\|_\varrho \ / \ \|u_n\|_{p,q,\varrho} = 0 \tag{11.6}$$

Next, we observe from (2.24), (Q-3), (1.11)(ii), (1.10), and (1.5)(iii) that there is a positive constant c_4 such that

$$c_4\mathcal{L}(u_n, u_n) \leq \mathcal{Q}(u_n, u_n) \ \ \forall n.$$

But then it follows from (11.5), (2.28), and this last inequality that

$$c_4\mathcal{L}(u_n, u_n) + \varepsilon_o^* \|u_n\|_\varrho^2 \leq$$

$$\{\int_\Omega \left|\tilde{h}_o\right|^{\theta^*} \varrho\}^{1/\theta^*} \{\int_\Omega |u_n|^\theta \varrho\}^{1/\theta} + \lambda_1 \|u_n\|_\varrho^2 + |G(u_n)| \quad \forall n.$$

On the other hand, we see from (4.2), (4.3), and this last inequality that there is a positive constant K_1 such that

$$\|u_n\|^2_{p,q,\varrho} \leq$$

$$K_1[\{\int_\Omega \left|\tilde{h}_o\right|^{\theta^*} \varrho\}^{1/\theta^*}\{\int_\Omega |u_n|^\theta \varrho\}^{1/\theta} + \lambda_1 \|u_n\|^2_\varrho + |G(u_n)|] \ \ \forall n.$$

Consequently, dividing both sides of this last inequality by $\|u_n\|^2_{p,q,\varrho}$ and passing to the limit as $n \to \infty$, we obtain from Lemma 7, the fact that $G \in [H^1_{p,q,\varrho}]'$, (11.4), and (11.6) that $1 \leq 0$. This is a manifest contradiction. Hence (11.4) does not hold, and (11.3) is therefore true.

Since $H^1_{p,q,\varrho}$ is a separable Hilbert space, we see from (11.3), Lemma 2, and Theorem 9 that there exists a subsequence (which for ease of notation we take to be a full sequence) and a function $u^* \in H^1_{p,q,\varrho}$ with the following properties:

$$\lim_{n\to\infty} [\ \|u_n - u^*\|_\varrho + \int_\Omega |\ u_n - u^*\ |^\theta \varrho\ ] = 0; \tag{11.7}$$

$$\exists w \in L^2_\varrho \cap L^\theta_\varrho \ s.t. \ |u_n(x)| \leq w(x) \quad \text{for a.e. } x \in \Omega, \ \ \forall n; \tag{11.8}$$

$$\lim_{n\to\infty} u_n(x) = u^*(x) \quad \text{for a.e.} x \in \Omega; \tag{11.9}$$

$$\lim_{n\to\infty} < D_i u_n, v >_{p_i} = < D_i u^*, v >_{p_i} \quad \forall v \in L^2_{p_i}, i = 1, \cdots, N; \tag{11.10}$$

$$\lim_{n\to\infty} < u_n, v >_q = < u^*, v >_q \quad \forall v \in L^2_q; \tag{11.11}$$

$$\lim_{n\to\infty} G(u_n) = G(u^*) \tag{11.12}$$

From this point the proof proceeds along the lines of the proof of Theorem 1, except that we have to take the superlinear assumption (f-8) into our reasoning. What we want to show is

$$\text{(5.25) holds.} \tag{11.13}$$

Since (f-8) does not enter into consideration, the exact same proof as before but using (11.7)-(11.12) shows that (11.13) will follow provided we show that

$$\text{(5.26) and (5.27) hold.} \tag{11.14}$$

An examination of the proofs of (5.26) and (5.27) shows that their proofs remain the same except for one place where (f-8) enters, namely

$$\lim_{n\to\infty} \mathcal{Q}(u_n, u_n - P_n u^*) = 0 \tag{11.15}$$

where $P_n u^*$ is the orthogonal projection u* into S_n.

We now show using (f-8), (11.3), (11.5), and (11.7)-11.12) that (11.15) holds. In particular from (11.5), we see that

$$\mathcal{Q}(u_n, u_n - P_n u^*) = (\lambda_1 - n^{-1}) < u_n, u_n - P_n u^* >_\varrho \tag{11.16}$$

$$+ \int_\Omega f(x, u_n)(u_n - P_n u^*)\varrho \; - G(u_n - P_n u^*).$$

Now from (4.3) and Lemma 1 we see that (5.38) holds. Consequently, from Lemma 7, we have that

$$\lim_{n \to \infty} [\| P_n u^* - u^* \|_\varrho + \int_\Omega |P_n u^* - u^*|^\theta \varrho] = 0$$

Hence, it follows from (11.7), Minkowski's inequality, and this last limit that

$$\lim_{n \to \infty} [\| u_n - P_n u^* \|_\varrho + \int_\Omega |u_n - P_n u^*|^\theta \varrho] = 0 \tag{11.17}$$

Also, (5.38) implies that $\lim_{n \to \infty} G(P_n u^* - u^*) = 0$. Therefore the linearity of G in conjunction with (11.12) and this last fact gives that

$$\lim_{n \to \infty} G(u_n, u_n - P_n u^*) = 0$$

As a consequence, we obtain from (11.16), (11.17), (11.3), and this last limit that (11.15) will hold provided we show

$$\lim_{n \to \infty} \int_\Omega f(x, u_n)(u_n - P_n u^*)\varrho = 0 \tag{11.18}$$

To establish (11.18), we observe from (f-8) that

$$|f(x, u_n)| \, |u_n - P_n u^*| \le K |u_n|^{\theta - 1} |u_n - P_n u^*| + h_o(x) |u_n - P_n u^*| \tag{11.19}$$

for $a.e. x \in \Omega$ where $h_o \in L_\varrho^{\theta^*}$, $\theta^* = \theta / (\theta - 1)$, and K is a positive constant. Now it follows from (11.7) that there is another constant K_1 such that

$$\int_\Omega |u_n|^{(\theta - 1)\theta^*} \varrho \le K_1 \;\; \forall n.$$

This fact in conjunction with (11.17) and (11.19) shows that (11.18) is indeed true. Hence (11.15) is true, and this in turn implies that (11.14) is true. We conclude that (11.13) is true and consequently that there exists a subsequence $\{u_{n_j}\}_{j=1}^{\infty}$ such that

$$\lim_{n\to\infty} Du_{n_j}(x) = Du^*(x) \text{ for a.e. } x \in \Omega. \tag{11.20}$$

Next, we let $v_J \in S_J$. Then it follows from (11.2) that for $n \geq J$

$$\mathcal{Q}(u_n, v_J) = (\lambda_1 - n^{-1}) < u_n, v_J >_\varrho + \int_\Omega f(x, u_n) v_J \varrho \; - G(v_J). \tag{11.21}$$

Now as in the proof of Theorem 1, using (11.3), (11.7) - (11.12), and (11.20), we obtain that

$$\lim_{n\to\infty} \mathcal{Q}(u_n, v_J) = \mathcal{Q}(u^*, v_J) \tag{11.22}$$

Next, we observe from (f-8) and (11.8) that

$$|f(x, u_n)| \leq K\,|w(x)|^{\theta-1} + h_o \text{ for a.e. } x \in \Omega,\ n = 1, 2, \cdots. \tag{11.23}$$

where $w \in L^\theta_\varrho$and $h_o \in L^{\theta^*}_\varrho$. From (f-1) and (11.9), we further obtain that

$$\lim_{n\to\infty} f(x, u_n(x)) = f(x, u^*(x)) \text{ for a.e. } x \in \Omega.$$

Now $|w|^{\theta-1} \in L^{\theta^*}_\varrho$. Hence we conclude from Holder's inequality, the Lebesgue dominated convergence theorem, Lemma 7, and (11.23) that

$$\lim_{n\to\infty} \int_\Omega f(x, u_n) v_J \varrho = \int_\Omega f(x, u^*) v_J \varrho. \tag{11.24}$$

But then passing to the limit on both sides of (11.21), we obtain from (11.22), (11.24), and (11.7) that

$$\mathcal{Q}(u^*, v_J) = \lambda_1 < u^*, v_J >_\varrho + \int_\Omega f(x, u^*) v_J \varrho \; - G(v_J) \tag{11.25}$$

Next given $v \in H^1_{p,q,\varrho}$, we replace v_J with $P_J v$ in (11.25). Since from (4.3) and Lemma 1, $\|P_J v - v\|_{p,q,\varrho} \to \infty$ as $J \to \infty$, it is then an easy matter to obtain from (11.25) and Lemma 7 that

$$\mathcal{Q}(u^*, v) = \lambda_1 < u^*, v >_\varrho + \int_\Omega f(x, u^*) v \varrho \; - G(v) \;\; \forall v \in H^1_{p,q,\varrho},$$

which fact concludes the proof of Theorem 10.

To prove Theorem 11, we set $f_n(x,s) = f(x,s) - s/n$. Then from (2.30), we see that $f_n(x,s)$ meets (2.28) with $\varepsilon_o^* = 1/n$. Hence we can invoke Lemma 8 and obtain that

$$\exists \{u_n\}_{n=2}^{\infty} \text{ with } u_n \in S_n \text{ such that} \tag{11.26}$$

$$\mathcal{Q}(u_n, v) = (\lambda_1 - 2/n) < u_n, v >_\varrho + \int_\Omega f(x, u_n) v\varrho \; - G(v) \;\; \forall v \in S_n$$

We claim there is a constant K such that (11.3) holds. Suppose to the contrary that (11.3) does not hold. Then we assume that (11.4) holds and show that this leads to a contradiction. Now the contradiction this time will not be so easy to reach since (2.28) is no longer valid for f(x,s). In the case we are now considering, the contradiction will involve the strict inequalities that occur in (2.31).

So assuming (11.26) and (11.4), we replace v in (11.26) with u_n and obtain

$$\mathcal{L}(u_n, u_n) = (\lambda_1 - 2/n) < u_n, u_n >_\varrho + \int_\Omega f(x, u_n) u_n \varrho \; - G(u_n) \tag{11.27}$$

$$+\mathcal{L}(u_n, u_n) - \mathcal{Q}(u_n, u_n)$$

Next, applying Lemma 1 and (2.30) to (11.27), we obtain

$$\sum_{j=2}^{n} (\lambda_j - \lambda_1) \left|\hat{u}_n(j)\right|^2 \leq \{\int_\Omega \left|\tilde{h}_o\right|^{\theta^*} \varrho\}^{1/\theta^*} \{\int_\Omega |u_n|^\theta \varrho\}^{1/\theta} + |G(u_n)| \tag{11.28}$$

$$+\mathcal{L}(u_n, u_n) - \mathcal{Q}(u_n, u_n).$$

We observe that there is a constant K_1 such that

$$|G(u_n)| \leq K_1 \left\|u_n\right\|_{p,q,\varrho} \;\; \forall n.$$

Also, we see from $(\mathrm{V}_L - 2)$ there is a constant $\varepsilon_2 > 0$ such that

$$\varepsilon_2 \lambda_j \leq \lambda_j - \lambda_1 \quad \forall j \geq 2.$$

Hence, we obtain from Lemma 7 and (11.28) that

$$\varepsilon_2 \sum_{j=2}^{n} \lambda_j \left|\hat{u}_n(j)\right|^2 \leq [\{\int_\Omega \left|\tilde{h}_o\right|^{\theta^*} \}^{1/\theta^*} K_\theta + K_1] \left\|u_n\right\|_{p,q,\varrho}$$

$$+\mathcal{L}(u_n, u_n) - \mathcal{Q}(u_n, u_n).$$

Letting $P_1 u_n = \hat{u}_n(1)\varphi_1$, applying Lemma 1 once again, and using the fact that Q is #-#-related to L (all we really need here is that Q is #-related to L), we conclude from this last inequality, since $\varepsilon_2 > 0$, that

$$\lim_{n\to\infty} \mathcal{L}\left(u_n - P_1 u_n, u_n - P_1 u_n\right) / \left\|u_n\right\|^2_{p,q,\varrho} = 0. \tag{11.29}$$

Now from Lemma 1 and $(V_L - 2)$, we also have that

$$\lambda_2 \left\|u_n\right\| \varrho^2 \leq \mathcal{L}\left(u_n - P_1 u_n, u_n - P_1 u_n\right) \quad \text{for } n \geq 2$$

where λ_2 is strictly positive. Hence it follows from (4.2), that $\mathcal{L}$ can be replaced by $\mathcal{L}_1$ in (11.29). But by (2.24) and (1.3),

$$\mathcal{L}_1\left(u_n - P_1 u_n, u_n - P_1 u_n\right) = \left\|u_n - P_1 u_n\right\|^2_{p,q,\varrho}$$

Therefore, we obtain from (11.29) that

$$\lim_{n\to\infty} \| \, u_n - P_1 u_n \, \|_{p,q,\varrho} / \left\|u_n\right\|_{p,q,\varrho} = 0. \tag{11.30}$$

Now λ_1 is an L-pseudo-eigenvalue of Q by hypothesis. Consequently, we infer from (11.4), (11.30), and (2.18) that

$$\lim_{n\to\infty} \left|\mathcal{Q}(u_n, P_1 u_n) - \mathcal{L}(u_n, P_1 u_n)\right| / \left\|u_n\right\|_{p,q,\varrho} = 0. \tag{11.31}$$

Next, we observe from (2.22) and the fact that Q is #-#-related to L that

$$\exists K_2 > 0 \text{ such that for n} \geq 2, \tag{11.32}$$

$$\left|\mathcal{Q}(u_n, u_n - P_1 u_n) - \mathcal{L}(u_n, u_n - P_1 u_n)\right| \leq K_2 \left\|u_n - P_1 u_n\right\|_{p,q,\varrho}.$$

But then it follows from (11.30) that

$$\lim_{n\to\infty} \left|\mathcal{Q}(u_n, u_n - P_1 u_n) - \mathcal{L}(u_n, u_n - P_1 u_n)\right| / \left\|u_n\right\|_{p,q,\varrho} = 0.$$

This fact joined with (11.31) then enables us to conclude that

$$\lim_{n\to\infty} \left|\mathcal{Q}(u_n, u_n) - \mathcal{L}(u_n, u_n)\right| / \left\|u_n\right\|_{p,q,\varrho} = 0 \tag{11.33}$$

Next, we recall from (1.3) that $< \cdot, \cdot >_{p,q,\varrho}$ is an inner product on $H^1_{p,q,\varrho}$. Therefore

$$\left\|u_n\right\|^2_{p,q,\varrho} = \left\|P_1 u_n\right\|^2_{p,q,\varrho} + \left\|u_n - P_1 u_n\right\|^2_{p,q,\varrho}.$$

Consequently, from (11.30) we obtain that

$$\lim_{n\to\infty} \|P_1 u_n\|_{p,q,\varrho} / \|u_n\|_{p,q,\varrho} = 1.$$

Now, $\mathcal{L}_1(P_1u_n, P_1u_n) = (1+\lambda_1)|\hat{u}_n(1)|^2 \leq (1+\lambda_1)\|u_n\|_\varrho^2$. But from (4.2) and (2.24), $\mathcal{L}_1(P_1u_n, P_1u_n) = \|P_1u_n\|_{p,q,\varrho}^2$. So we infer from this last limit that

$$\exists K_3 > 0 \text{ such that} \tag{11.34}$$

$$\|u_n\|_{p,q,\varrho} \leq K_3 \|u_n\|_\varrho \quad \forall n.$$

Continuing with the proof to show that (11.4) leads to a contradiction, we set

$$v_n = u_n / \|u_n\|_\varrho . \tag{11.35}$$

and

$$v_{1n} = \hat{v}_n(1)\varphi_1 \text{ and } v_{2n} = v_n - v_{1n} \tag{11.36}$$

From (11.34) and (11.35), we see that $\|v_n\|_{p,q,\varrho} \leq K_3 \;\; \forall n$. Consequently, we have from Lemma 2 and Lemma 7 that there exists $v^* \in H^1_{p,q,\varrho}$ and a subsequence (which for ease of notation we take to be the full sequence) such that

$$\lim_{n\to\infty}\left[\|v_n - v^*\|_\varrho + \int_\Omega |v_n - v^*|^\theta \varrho\right] = 0; \tag{11.37}$$

$$\exists w \in L^2_\varrho \cap L^\theta_\varrho \text{ s.t. } |v_n(x)| \leq w(x) \text{ for a.e. } x \in \Omega, \forall n; \tag{11.38}$$

$$\lim_{n\to\infty} v_n(x) = v^*(x) \quad \text{for a.e. } x \in \Omega; \tag{11.39}$$

$$\lim_{n\to\infty} < D_i v_n, V >_{p_i} = < D_i v^*, V >_{p_i} \;\; \forall V \in L^2_{p_i}; \tag{11.40}$$

$$\lim_{n\to\infty} < v_n, V >_q = < v^*, V >_q \;\; \forall V \in L^2_q; \tag{11.41}$$

$$\lim_{n\to\infty} G(v_n) = G(v^*). \tag{11.42}$$

Also from (11.30) and (11.34)-(11.36), we see that $\lim_{n\to\infty}\|v_{2n}\|_\varrho = 0$ and that $1 = \|v_{1n}\|_\varrho^2 + \|v_{2n}\|_\varrho^2$. Consequently $\lim_{n\to\infty}\|v_{1n}\|_\varrho^2 = 1$ and therefore

$$\lim_{n\to\infty} |\hat{v}_n(1)| = 1 \text{ and } \lim_{n\to\infty} |\hat{v}_n(j)| = 0 \;\text{ for } j \geq 2. \tag{11.43}$$

From (11.37), we also have that $\lim_{n\to\infty} |\hat{v}_n(j) - \hat{v}_n^*(j)| = 0$ for $j \geq 1$. We consequently conclude from (11.43) that

$$v^* = \hat{v}^*(1)\varphi_1 \text{ where } |\hat{v}^*(1)| = 1. \tag{11.44}$$

We shall assume

$$\hat{v}^*(1) = 1 \text{ and } v^* = \varphi_1 \tag{11.45}$$

and arrive at a contradiction. Similar reasoning will lead to a contradiction for the assumption $\hat{v}^*(1) = -1$ and $v^* = -\varphi_1$.

To obtain this contradiction, we first observe from ($V_L - 2$), (11.4), (11.34), (11.39), and (11.45) that $\lim_{n\to\infty} u_n(x) = +\infty$ for a.e. $x \in \Omega$. Next, we see from ($V_L - 2$) once again in conjunction with Lemma 1 that

$$\mathcal{L}(u_n, u_n) - \lambda_1 < u_n, u_n >_\varrho \geq 0 \quad \forall n.$$

Consequently, it follows from (11.27), (11.33), and (11.34) that given $\varepsilon > 0$ there is a n_o such that for $n \geq n_o$.

$$G(v_n) \leq \int_\Omega f(x, u_n)v_n\varrho + \varepsilon \quad \text{for } n \geq n_o. \tag{11.46}$$

But from (2.30) and (11.38)

$$f(x, u_n)v_n \leq h_o(x)w(x) \quad \text{for a.e. x} \in \Omega \text{ and } \forall n.$$

Also, $h_o \in L_\varrho^{\theta^*}$ and $w \in L_\varrho^\theta$. Therefore $h_o w \in L_\varrho^1$, and we conclude from Fatou's Lemma (See [Rud, p.24].) in conjunction with (11.39), (11.42), and (11.45) applied to (11.46) that

$$G(\varphi_1) \leq \int_\Omega f^+(x)\varphi_1\varrho + \varepsilon$$

Since $\varepsilon > 0$ is arbitrary, we conclude from this last inequality that

$$G(\varphi_1) \leq \int_\Omega f^+(x)\varphi_1\varrho.$$

But this last inequality is a direct contradiction to the first inequality in (2.31). A similar contradiction to the second inequality in (2.31) would have occurred if in (11.44) we would have chosen $\hat{v}^*(1) = -1$ and $v^* = -\varphi_1$. Hence (11.4) is false, and (11.3) is indeed true.

Once (11.3) holds the rest of the proof of Theorem 11 proceeds exactly along the same lines as those of Theorem 10 because wherever we use (2.28) in this last part of the proof it turns out (2.30) will suffice. Hence the proof Theorem 11 is complete.

To prove Theorem 12 we obtain just as in Theorem 11 that (11.26) holds. What we want to show in this proof first is that (11.3) is valid. So we assume that (11.4) holds and will show that this assumption leads to a contradiction of the inequalities in (2.31). Once again as in the proof of Theorem 11 we obtain (11.27) and (11.28) and conclude from the fact that Q is #-related to L that both (11.29) and (11.30) hold. At this point in the proof of Theorem 11 the notion of λ_1 being an L-pseudo-eigenvalue kicks in. So to obtain a contradiction using (11.4), the proof Theorem 12 becomes different from Theorem 11 after (11.30).

From (11.30), we have in particular with $P_1 u_n = \hat{u}_n(1)\varphi_1$ that

$$\lim_{n\to\infty} \|P_1 u_n\|_{p,q,\varrho} / \|u_n\|_{p,q,\varrho} = 1$$

Now as we have seen before from (2.24) and (4.2),

$$\|P_1 u_n\|_{p,q,\varrho} = (1+\lambda_1)\,|\hat{u}_n(1)|^2 \leq (1+\lambda_1)\,\|u_n\|_\varrho .$$

So, it follows from this last limit that there is a positive constant K_4 such that

$$\lim_{n\to\infty} \inf \| u_n \|_\varrho / \|u_n\|_{p,q,\varrho} \geq K_4 > 0$$

Consequently

$$\exists K_5 > 0 \text{ such that}$$

$$\|u_n\|_{p,q,\varrho} \leq K_5 \|u_n\|_\varrho \quad \forall n. \tag{11.47}$$

Next, we return to (11.26) and obtain that

$$\mathcal{L}(u_n,u_n) - \lambda_1 < u_n, u_n >_\varrho + \mathcal{Q}(u_n,u_n) - \mathcal{L}(u_n,u_n) \tag{11.48}$$

$$\leq \int_\Omega f(x,u_n)u_n\varrho - G(u_n)$$

Now from Lemma 1, we have that

$$\mathcal{L}(u_n,u_n) - \lambda_1 < u_n, u_n >_\varrho \geq 0.$$

Also, from (2.32) and (11.47), we see that given $\varepsilon > 0$

$$\mathcal{Q}(u_n,u_n) - \mathcal{L}(u_n,u_n) \geq -\varepsilon \|u_n\|_\varrho \quad \text{for } n \geq n_o.$$

So given $\varepsilon > 0$, we obtain from these last two inequalities applied to (11.48) that

$$\exists n_o > 0 \text{ such that for } n \geq n_o \tag{11.49}$$

$$G(u_n) \leq \int_\Omega f(x, u_n) u_n \varrho + \varepsilon \|u_n\|_\varrho .$$

On dividing both sides of (11.49) by $\|u_n\|_\varrho$, we obtain exactly (11.46). Since it is (11.30) (which is also valid for this proof of Theorem 12) in conjunction with (11.47) which gives us (11.43), we now can assume (11.45) in conjunction with (11.46) and the rest of the proof of Theorem 12 proceeds exactly and precisely as the proof of Theorem 11. Hence the proof of Theorem 12 is complete.

Chapter 2

Quasilinear Parabolic Equations

2.1 Introduction

In this chapter, we present two theorems, based on the previous developments in Chapter 1, for time-periodic singular quasilinear parabolic differential equations. The first will be the time-periodic analogue of the double resonance result (i.e., Theorem 4) of Chapter 1, and the second will deal with a reaction-diffusion time-periodic system and will constitute a far-reaching analogue of Theorem 6 of Chapter 1. In particular, this latter theorem will make use of the notion of L-pseudo-eigenvalues first introduced in the previous chapter. Both of the theorems in this chapter appear to be new even in the singular semilinear case.

The results in this chapter are motivated by the author's previous joint papers with Lefton [LeS] and Legner [LgS]. These two papers in turn were motivated by the time- periodic work of Brezis and Nirenberg [BN, Ch. V], Castro and Lazer [CL], and Mckenna and Walter [MW].

With the point of view of time-periodicity in mind, we leave $\Omega \subset \mathbf{R}^N$, $\Gamma \subset \partial\Omega$, $\varrho(x)$ and $p_i(x)$ be as in Chapter 1 with ϱ, $p_i \in C^{\,0}(\Omega)$, positive functions satisfying (1,1.1) for $i = 1, ..., N$. On the other hand, $q(x)$ will satisfy the inequality

$$\exists K > 0 \; s.t. \; 0 \leq q(x) \leq K\varrho(x) \qquad \forall x \in \Omega. \tag{1.1}$$

We designate by $\mathcal{A}$ the set of real-valued functions defined as follows:

$$\mathcal{A} = \{u : u \in C^0\left(\bar{\Omega} \times \mathbf{R}\right), \; u(x, t + 2\pi) = u(x, t) \;\; \forall (x, t) \in \bar{\mathbf{\Omega}} \times \mathbf{R}\},$$

and introduce the pre-Hilbert space:

$$\tilde{C}^1_{p,\varrho}(\Omega,\Gamma) = \{u : u \in \mathcal{A} \cap C^{\,2}(\Omega \times \mathbf{R}),\ u(x,t) = 0$$

$$\forall (x,t) \in \ \Gamma \times \mathbf{R}, \int_{\tilde{\Omega}} \left[\sum_{i=1}^{N} p_i \mid D_i u \mid^2 + \varrho(u^2 + |D_t u|^2) \right] < \infty\}$$

where $\tilde{\Omega} = \Omega \times T$, $T = (-\pi, \pi)$ and $p = (p_1, ..., p_N)$. (In the sequel, we shall write $\tilde{C}^1_{p,\varrho}$ for $\tilde{C}^1_{p,\varrho}(\Omega,\Gamma)$.)

We designate the inner product in $\tilde{L}^2_\varrho = L^2_\varrho\left(\tilde{\Omega}\right)$ by

$$< u, v >_{\tilde{\varrho}} = \int_{\tilde{\Omega}} u(x,t)v(x,t)\varrho(x)\ dxdt.$$

In a similar manner, we designate the corresponding inner products in $\tilde{L}^2_{p_i}$ and $\tilde{L}^2_q$ by $< u, v >_{\tilde{p_i}}$ and $< u, v >_{\tilde{q}}$. In the pre-Hilbert space $\tilde{C}^1_{p,\varrho}$, we then introduce the inner product as follows:

$$< u, v >_{\tilde{H}} = \sum_{i=1}^{N} < D_i u, D_i v >_{\tilde{p_i}} + < u, v >_{\tilde{\varrho}} + < D_t u, D_t v >_{\tilde{\varrho}} \qquad (1.2)$$

The Hilbert space that is generated by completing $\tilde{C}^1_{p,\varrho}$ by the method of Cauchy sequences with respect to the norm generated by the inner product in (1.2), we call $\tilde{H} = \tilde{H}(\Omega, \Gamma)$.

Next we assume $a_0(x)$ and $a_{ij}(x)$ meet $(1, 1.5)\,(i) - (iv)$ in Ω and we define Lu in Ω as before by

$$Lu = -\sum_{i,j=1}^{N} D_i[p_i^{\frac{1}{2}} p_j^{\frac{1}{2}} a_{ij} D_j u] + a_0 q u. \qquad (1.3)$$

The corresponding two-form on Ω then is

$$\mathcal{L}(u,v) = \sum_{i,j=1}^{N} \int_\Omega p_i^{\frac{1}{2}} p_j^{\frac{1}{2}} a_{ij} D_j u D_i v + \int_\Omega a_0 u v q \qquad (1.4)$$

$$\text{for } u, v \in H^1_{p,q,\varrho}(\Omega, \Gamma)\ ,$$

as defined in Chapter 1.

In the sequel, we shall suppose that (Ω, Γ) is a V_L–region (as outlined in Chapter 1); so in particular $(V_L - 1)$ and $(V_L - 2)$ of Chapter 1 hold for the L defined in (1.3) above. We note in passing that if $\varphi(x) \in C^1_{p,q,\varrho}(\Omega,\Gamma)$, which is defined in (1,1.2) with q meeting (1.1) above, then $\varphi(x) \in \tilde{C}^1_{p,\varrho}$. Hence it follows that $\varphi_0(x), \varphi_n(x), \varphi_n(x)\ cos\ jt$, and $\varphi_n(x)\ sin\ jt$ are in $\tilde{H} = \tilde{H}(\Omega,\Gamma)$, as just defined above, for n and j positive integers where the φ_n are the λ_n–eigenfunctions described in $(V_L - 1)$.

We extend the bilinear form $\mathcal{L}(\cdot,\cdot)$ defined in (1.3) on functions in $\tilde{H}$ by

$$\mathcal{L}^{\sim}(u,v) = \int_T \mathcal{L}(u,v) = \sum_{i,j=1}^{N} \int_{\tilde{\Omega}} p_i^{\frac{1}{2}} p_j^{\frac{1}{2}} a_{ij} D_j u D_i v + < a_0 u, v >_q^{\tilde{}} \tag{1.5}$$

where $< \cdot,\cdot >_q^{\tilde{}}$ *is* understood to be zero in case $q \equiv 0$.

Next, we introduce the quasilinear Q which we shall be dealing with in this chapter. In particular

$$Qu = \sum_{i=1}^{N} D_i[p_i^{\frac{1}{2}} A_i(x,u,Du)] + qB_0(x,u,Du)u, \tag{1.6}$$

and we make the following assumptions concerning $A_i (i = 1,...,N)$ and B_o :

$(Q-1),(Q-2),(Q-3), and (Q-4)$will be exactly as in Chapter1;

$$(Q-5)\ (i)\ B_0(x,s,\xi) : \Omega \times \mathbf{R} \times \mathbf{R}^N \to \mathbf{R}, \tag{1.7}$$

$$(ii)\ B_0(x,s,\xi) \in C^0(\Omega \times \mathbf{R} \times \mathbf{R}^N) \cap L^\infty(\Omega \times \mathbf{R} \times \mathbf{R}^N)$$

$$(iii)\ B_0(x,s,\xi) \geq 0 \quad \forall x \in \Omega, s \in \mathbf{R}, \text{and } \xi \in \mathbf{R}^N;$$

$$(Q-6) \int_{\tilde{\Omega}} [\sum_{i=1}^{N} p_i^{\frac{1}{2}} A(x,v,Dv)\, D_i D_t v + qB_o(x,v,Dv) v D_t v] = 0$$

$$\forall v \in \tilde{C}^{1\flat}_{p,\varrho} \ ,$$

where

$$\tilde{C}^{1\flat}_{p,\varrho} = \{v : v \in \tilde{C}^1_{p,\varrho} \text{ and } D_t v \in \tilde{C}^1_{p,\varrho}\}. \tag{1.8}$$

Assuming that Q given by (1.5) meets (Q-1)-(Q-5), we introduce the two-form

$$\mathcal{Q}^{\tilde{}}(u,v) = \sum_{i=1}^{N} \int_{\tilde{\Omega}} A_i(x,u,Du)\, D_i v\ p_i^{\frac{1}{2}} + \int_{\tilde{\Omega}} B_o(x,u,Du) uvq \tag{1.9}$$

for $u, v \in \tilde{H}$. We shall say Q given by (1.5) is $\#\tilde{H}$-related to L if the following two facts hold:

$$(i) \lim_{\|u\|_{\tilde{H}} \to \infty} [\mathcal{Q}^{\tilde{}}(u,v) - \mathcal{L}^{\tilde{}}(u,v)] / \|u\|_{\tilde{H}} = 0 \quad \text{uniformly for } \|v\|_{\tilde{H}} \leq 1; \tag{1.10}$$

(ii) $\exists\ c_4 > 0$ such that $q(x)\ B_0(x, u, Du) \geq c_4\ q(x)\ a_0(x)$

$\forall (x,t) \in \tilde{\Omega}$ and $\forall u \in \tilde{C}^1_{p,\varrho}$.

We see from (1.9) that if Q meets (Q-1)-(Q-5) then Q meets (Q-6) if furthermore

$$\mathcal{Q}^{\sim}(v, D_t v) = 0 \ \ \forall v \in \tilde{C}^{1b}_{p,\varrho}. \tag{1.11}$$

It appears that (Q-6) was introduced for the first time in [LfS], and this concept plays a key role in this chapter. That L given in (1.3) satisfies (Q-6), i.e., $\mathcal{L}^{\sim}(v, D_t v) = 0\ \ \forall v \in \tilde{C}^{1b}_{p,\varrho}$, follows from the fact that

$$2\mathcal{L}^{\sim}(v, D_t v) = \sum_{i,j=1}^{N} \int_{\tilde{\Omega}} a_{ij} p_i^{\frac{1}{2}} p_j^{\frac{1}{2}} (D_i v D_j D_t v + D_j v D_i D_t v)$$

$$= \sum_{i,j=1}^{N} \int_{\Omega} a_{ij} p_i^{\frac{1}{2}} p_j^{\frac{1}{2}} \int_{T} D_t (D_i v D_j v) = 0$$

2.2 Statement of main results

The first theorem in this chapter that we shall present for Q satisfying the $\#\tilde{H}$-relationship will be a double resonance theorem which is the analog of Theorem 4 in Chapter 1 and deals with the equation

$$\varrho D_t u + Qu = [\lambda_{j_o} u + f(x,u) + g(x,t,u)]\varrho \tag{2.1}$$

where λ_{j_o} is an eigenvalue of L of multiplicity j_1 (as described in Chapter 1); so $\lambda_{j_o + j_1}$ is the next eigenvalue strictly greater than λ_{j_o}. We shall suppose that f meets (f-1) and (f-2) of Chapter 1 where the γ in (f-2) is defined by (1,2.10). For the g defined in (2.1) above, we make the following assumptions:

(g-1) g(x,t,s) satisfies the usual Caratheodory assumptions on $\tilde{\Omega}$ (i.e., it is measurable in (x,t) for every fixed $s \in \mathbf{R}$, and continuous in s for a.e. fixed $(x,t) \in \tilde{\Omega}$);

(g-2) $\forall \varepsilon > 0, \exists g_\varepsilon \in \tilde{L}^2_\varrho$ s.t. $|g(x,t,s)| \leq \varepsilon |s| + g_\varepsilon(x,t)$
for *a.e.* $(x,t) \in \tilde{\Omega}$ and $\forall s \in \mathbf{R}$.

In §5 below, we prove the following theorem.

Theorem 1 *Let $\tilde{\Omega} = \Omega \times T, \varrho, p = (p_1, ..., p_N)$, and q be as described in §2.1. Let L and Q be given by (1.3) and (1.6) respectively and assume (1,1.5), (Q-1)-(Q-6), that Q is $\#\tilde{H}$-related to L, and that (Ω, Γ) is a V_L−region. Suppose also that λ_{j_o} is an eigenvalue of L of multiplicity j_1, that f and g satisfy (f-1), (f-2), and (g-1), (g-2) respectively. Assume furthermore that (1, 2.12) and (1, 2.13) hold for $\mathcal{F}^\pm$ and $\mathcal{F}_\pm$ which are defined in (1, 2.11). Then $\exists\ u^* \in \tilde{H}$ which is a weak solution of (2.1) on $\tilde{\Omega}$.*

To be quite explicit, what we mean by $u^* \in \tilde{H}$ is a weak solution of (2.1) on $\tilde{\Omega}$ is the following:

$$< D_t u^*, v >_\varrho^{\tilde{}} + \mathcal{Q}^{\tilde{}}(u^*, v) = \lambda_{j_o} < u^*, v >_\varrho^{\tilde{}} + < f(\cdot, u^*) + g(\cdot, \cdot, u^*), v >_\varrho^{\tilde{}} \tag{2.2}$$

$\forall v \in \tilde{H}$ where $\mathcal{Q}^{\tilde{}}$ is given by (1.8).

The second theorem that we prove in this chapter is in part a systems analogue of Theorem 6 of Chapter 1. This last quoted theorem is a resonance result but since we will be dealing with systems, we shall also incorporate some nonresonance aspects; so the theorem that we shortly state is both a nonresonance as well as a resonance result. The resonance part will also deal with the analogue of the notion of L-pseudo-eigenvalues introduced in Chapter I which we will now define for the parabolic situation.

First of all, we need the notion of #-#H̃. In particular, we say the operator Q given in (1.6) is #-#H̃ related to L if the following obtains:

$$(i) \text{ Q is } \#\tilde{H} - \text{ related to L}, \tag{2.3}$$

$(ii)\exists$ positive constants K_1 and K_2 such that.

$$\left|\mathcal{Q}^{\tilde{}}(u,v) - \mathcal{L}^{\tilde{}}(u,v)\right| \leq K_1 \left|\mathcal{L}^{\tilde{}}(v,v)\right|^{\frac{1}{2}} + K_2 \quad \forall u, v \in \tilde{H}.$$

Also, we shall say φ is an $\tilde{H}-\lambda_{j_o}$–eigenfunction of Q if

$$\mathcal{Q}^{\tilde{}}(v,\varphi)=\lambda_{j_o}\langle v,\varphi\rangle_{\varrho}^{\tilde{}}\qquad \forall v\in\tilde{H}$$

We next define the notion of an H̃-L-pseudo-eigenvalue. To do this, it will be convenient at this point to introduce a symbol for the absolute value on the right-hand side of the inequality in (2.3)(ii). In particular, we shall use the notation

$$\|v\|_{\mathcal{L}^{\tilde{}}}=\left|\mathcal{L}^{\tilde{}}(v,v)\right|^{\frac{1}{2}}\qquad \forall v\in\tilde{H},\qquad\text{and}\tag{2.4}$$

$$\|u\|_{\tilde{L}^2_{\varrho}}=[<u,u>_{\varrho}^{\tilde{}}]^{\frac{1}{2}}\quad\forall u\in\tilde{L}^2_{\varrho}=L^2_{\varrho}\left(\tilde{\Omega}\right).$$

Also, $S_n^{\ddagger}$ will stand for the subspace of $\tilde{H}$ introduced in (4.23) below and $\{\tilde{\varphi}^c_{nk}\}_{k=0,n=1}^{\infty\ ,\infty}$ the sequence introduced in (4.2)(i) below.

We say λ_{j_o} is an H̃-L-pseudo-eigenvalue of Q̃ provided the following obtains:

$(i)\lambda_{j_o}$ is an eigenvalue of L of multiplicity $j_1$with $\tilde{\varphi}^c_{j_o0},\ldots,\tilde{\varphi}^c_{j_o+j_1-1\ 0}$ the corresponding orthonormal eigenfunctions;

$(ii)P^o_{j_o}(u)$ is the projection of u onto the subspace of $\tilde{H}$ spanned by $\tilde{\varphi}^c_{j_o0},\ldots,\tilde{\varphi}^c_{j_o+j_1-1\ 0}$;

(iii) if $\{u_n\}_{n=1}^{\infty}$is a sequence with $u_n\in S_n^{\ddagger}$, $\|u_n\|_{\tilde{L}^2_{\varrho}}+\|u_n\|_{\mathcal{L}^{\tilde{}}}\to\infty$, and $[\left\|u_n-P^o_{j_o}(u_n)\right\|_{\tilde{L}^2_{\varrho}}+\left\|u_n-P^o_{j_o}(u_n)\right\|_{\mathcal{L}^{\tilde{}}}]/[\|u_n\|_{\tilde{L}^2_{\varrho}}+\|u_n\|_{\mathcal{L}^{\tilde{}}}]\to 0$, then

$$\lim_{n\to\infty}[\mathcal{Q}^{\tilde{}}(u_n,P^o_{j_o}u_n)-\mathcal{L}^{\tilde{}}(u_n,P^o_{j_o}u_n)]/[\|u_n\|_{\tilde{L}^2_{\varrho}}+\|u_n\|_{\mathcal{L}^{\tilde{}}}]=0.\tag{2.5}$$

It is clear that if λ_{j_o} is an eigenvalue of L, and every λ_{j_o} eigenfunction of L is also an $\tilde{H}-\lambda_{j_o}$–eigenfunction of Q, then λ_{j_o} is also an H̃-L-pseudo-eigenvalue of Q.

We shall study a quasilinear parabolic system consisting of $N_1+N_2+N_3$ equations of the following nature.

$$\varrho D_t u_i+\mu_i Q_i u_i=[\mu_i\lambda_{j_o(i)}u_i+f_i(x,t,u_i)\tag{2.6}$$

$$+g_i(x,t,u_1,\ldots,u_{N_1+N_2+N_3})-h_i\,]\,\varrho\qquad i=1,\ldots,N_1$$

$$=\left[\mu_i(\lambda_{j_o(i)}+\varepsilon_i)u_i+f_i(x,t,u_i)+g_i(x,t,u_1,\ldots,u_{N_1+N_2+N_3})-h_i\right]\varrho$$

$$i = N_1 + 1, \ldots, N_1 + N_2$$

$$= [\mu_i(\lambda_1 - \varepsilon_i)u_i + f_i(x,t,u_i) + g_i(x,t,u_1,\ldots,u_{N_1+N_2+N_3}) - h_i]\,\varrho$$

$$i = N_1 + N_2 + 1, \ldots, N_1 + N_2 + N_3.$$

In the above system, we are given a fixed L of the form (1.3), meeting (1,1.5)(i)-(iv). Also, we assume in the above system that each Q_i is of the form (1.6) and meets (Q-1)-(Q-6) as described in §2.1 of this chapter. In addition , we assume that each Q_i is #-#$\tilde{H}$ related to L for $i = 1, \ldots, N_1 + N_2 + N_3$. Furthermore, we assume that μ_i is a positive diffusion coefficient for $i = 1, \ldots, N_1 + N_2 + N_3$.

The $\lambda_{j_o(i)}$ for $i = 1, \ldots, N_1 + N_2$ will be eigenvalues of L of multiplicity $j_1(i)$. λ_1 will designate the first eigenvalue of L which by $(V_L - 1)$ in Chapter I is assumed to be simple. For $i = 1, \ldots, N_1$,$\lambda_{j_o(i)}$ is assumed to be an $\tilde{H}$-L-pseudo-eigenvalue for Q_i, and $\varepsilon_i > 0$ for $i = N_1 + 1, \ldots, N_1 + N_2 + N_3$. Also, with $\lambda_{j_o(i)+j_1(i)}$ designating the first eigenvalue of L strictly greater than $\lambda_{j_o(i)}$, we shall assume

$$0 < \varepsilon_i < [\,\lambda_{j_o(i)+j_1(i)} - \lambda_{j_o(i)}] \quad \text{for } i = N_1 + 1, \ldots, N_1 + N_2. \tag{2.7}$$

For f_i in the system (2.6) we shall assume the following:

(f-9) $f_i(x,t,s) : \tilde{\Omega} \times \mathbf{R} \to \mathbf{R}$ satisfies the Caratheodory conditions;
(f-10) $\exists\, c(x,t) \in L^2_\varrho(\tilde{\Omega})$ s.t. for $i = 1, \ldots, N_1 + N_2 + N_3$,

$$|\,f_i(x,t,s)| \le c(x,t) \quad \forall s \in \mathbf{R} \text{ and } a.e.\ (x,t) \in \tilde{\Omega}.$$

For f_i, $i = 1, \ldots, N_1$, in (2.6) we shall also use the following notation:

$$f_{i\pm}(x,t) = \lim_{s\to\pm\infty}\inf f_i(x,t,s) \text{ and } f_i^{\pm}(x,t) = \lim_{s\to\pm\infty}\sup f_i(x,t,s).$$

For g_i in the system (2.6) we shall assume the following:

(g-3) $g(x,t,s_1,\ldots,s_{N_1+N_2+N_3}) : \tilde{\Omega} \times \mathbf{R}^{N_1+N_2+N_3} \to \mathbf{R}$ is in $C^o(\tilde{\Omega} \times \mathbf{R}^{N_1+N_2+N_3})$;
(g-4) $\exists\, c^*(x,t) \in L^2_\varrho(\tilde{\Omega})$ s.t.

$$|g(x,t,s_1,\ldots,s_{N_1+N_2+N_3})| \leq c^*(x,t)$$

$$\forall (s_1,\ldots,s_{N_1+N_2+N_3}) \in \mathbf{R}^{N_1+N_2+N_3} \quad \text{and } (x,t) \in \tilde{\Omega};$$

(g-5)

$$\lim_{|s_i|\to\infty} g_i(x,t,s_1,\ldots,s_{i-1},s_i,s_{i+1},s_{N_1+N_2+N_3}) = 0$$

uniformly for $(x,t) \in \tilde{\Omega}$ and $\mathrm{s}_j \in \mathbf{R}$, $j = 1,\ldots,i-1,i+1,\ldots,N_1+N_2+N_3$.

For the time-periodic system (2.6), the theorem that we establish is the following:

Theorem 2   *Let $\tilde{\Omega} = \Omega \times T$, ϱ, $p = (p_1,\ldots,p_N)$, and q be as described in §2.1. Also, let L and Q_i be given by (1.3) and (1.6) respectively and suppose that μ_i and $\lambda_{j_o(i)}$ are as described above for $i = 1,\ldots,N_1+N_2+N_3$ with $\lambda_{j_o(i)}$ an $\tilde{H}$-L-pseudo-eigenvalue of Q_i for $i = 1,\ldots,N_1$. Let $\varepsilon_i > 0$ for $i = N_1+1,\ldots,N_1+N_2+N_3$ and meet (2.7) for $i = N_1+1,\ldots,N_1+N_2$. Suppose furthermore that f_i meets (f-9) and (f-10), that g_i meets (g-3) -(g-5), that $h_i \in L^2_\varrho(\tilde{\Omega})$, and that Q_i is #-#$\tilde{H}$ related to L for $i = 1,\ldots,N_1+N_2+N_3$. Assume, also, for $i = 1,\ldots,N_1$ that*

$$\int_{\tilde{\Omega}} h_i(x,t)w\varrho < \int_{\tilde{\Omega}\cap[w>0]} f_{i+}(x,t)w\varrho + \int_{\tilde{\Omega}\cap[w<0]} f_i^-(x,t)w\varrho \qquad (2.8)$$

for every $w(x)$ which is a nontrivial $\lambda_{j_o(i)}$-eigenfunction of L. Then $\exists$ $(u_1^*,\ldots,u_{N_1+N_2+N_3}^*) \in \tilde{H}^{N_1+N_2+N_3}$ which is a weak solution of the system in (2.6) on $\tilde{\Omega}$.

To be specific, what is meant by the last statement in the theorem is the following: $\exists$ $u_i^* \in \tilde{H}$ for $i = 1,\ldots,N_1+N_2+N_3$ such that

$$< D_t u_i^*, v >_\varrho^{\sim} + \mu_i \tilde{Q_i}(u,v) = \qquad (2.9)$$

$$\mu_i \lambda_{j_o(i)} < u_i^*, v >_\varrho^{\sim} + < f_i(\cdot,\cdot,u_i^*) + g_i(\cdot,\cdot,u_1^*,\ldots,u_{N_1+N_2+N_3}^*)$$

$$-h_i(\cdot,\cdot), v >_\varrho^{\sim} \quad \forall v \in \tilde{H}, \qquad i = 1,\ldots,N_1,$$

$$\mu_i(\lambda_{j_o(i)} + \varepsilon_i < u_i^*, v >_\varrho^{\sim} + < f_i(\cdot,\cdot,u_i^*) + g_i(\cdot,\cdot,u_1^*,\ldots,u_{N_1+N_2+N_3}^*)$$

$$-h_i(\cdot,\cdot), v >_{\varrho}^{\tilde{}} \quad \forall v \in \tilde{H},\ i = N_1+1, \ldots, N_1+N_2,$$

$$\mu_i(\lambda_1 - \varepsilon_i < u_i^*, v >_{\varrho}^{\tilde{}} + < f_i(\cdot,\cdot,u_i^*) + g_i(\cdot,\cdot,u_1^*, \ldots, u_{N_1+N_2+N_3}^*)$$

$$-h_i(\cdot,\cdot), v >_{\varrho}^{\tilde{}} \quad \forall v \in \tilde{H}, \quad i = N_1+N_2+1, \ldots, N_1+N_2+N_3.$$

Remark 1 *The theorem above is still valid in case the part of the system from $N_1+1, \ldots, N_1+N_2+N_3$ is missing. Likewise if the part of the system from $1, \ldots, N_1$ is missing the theorem is still valid without (2.8). A similar statement prevails in case other parts of the system are missing.*

To give some insight into the motivation for the two theorems in this chapter, with $\tilde{\Omega} = \Omega \times T$, we focus on the equation

$$\varrho D_t u + Qu = [\lambda u + f(x,t,u) - h(x,t)]\varrho \tag{2.10}$$

where $\lambda \in \mathbf{R}$, $f \in C^0(\tilde{\Omega} \times \mathbf{R}) \cap L^\infty(\tilde{\Omega} \times \mathbf{R})$, h$\in C^0(\tilde{\Omega}) \cap L^\infty(\tilde{\Omega})$, and Q is a quasilinear elliptic operator meeting the axioms (Q-1)-(Q-6) stated in section one of this chapter. In [LfS] it is established that the equation in (2.10) always has a weak solution in $\tilde{\mathrm{H}}$ for $\lambda < \lambda_1^*$ provided $\mathrm{p}_i = \varrho = 1$, q=0, and we are working with the Dirichlet problem where λ_1^* is defined as in (1,2.34) (except we now use u$\in\tilde{\mathrm{H}}$ in the definition of λ_1^*). Also, it is shown in [LfS] if Q is asymptotically related to a linear elliptic operator L, then a resonance result (once again for the Dirichlet problem) can be obtained for the equation in (2.10) for $\lambda = \lambda_1^*$. The key new ingredient for these quasilinear results is (Q-6) which evidently was used for the first time in [LfS].

Results were therefore known for the equation in (2.10) when $\lambda \leq \lambda_1^*$. The question then becomes : *How does one proceed when* $\lambda > \lambda_1^*$? Answers to this question are given by Theorems 1&2. In particular, we see from Theorem 1 and the technique in Remark 2 (which appears after Theorem 4 in Chapter 1) that if Q is #$\tilde{\mathrm{H}}$-related to L and $\lambda_{j_o} < \lambda < \lambda_{j_o+j_1}$ (where λ_{j_o} and $\lambda_{j_o+j_1}$ are successive eigenvalues of L), then a weak solution u* $\in \tilde{\mathrm{H}}$ always exist for the equation in (2.10). Also, in case $\lambda = \lambda_{j_o}$ is an $\tilde{\mathrm{H}}$-L-pseudo-eigenvalue of Q (with Q now being #-#$\tilde{\mathrm{H}}$-related to L) and the criteria given in (2.8) for i=1 is met , then a solution u* $\in \tilde{\mathrm{H}}$ to the equation in (2.10) exists. Since $\lambda_1^* = \lambda_1$ in this case, we see that the equation in (2.10) is covered by Theorems 1&2 for $\lambda > \lambda_1$, and hence for *all* $\lambda \in \mathbf{R}$.

Theorem 2 above, dealing with time-periodic reaction-diffusion equations, is motivated by the material contained in [LgS] and covers similar results to those just mentioned in the previous paragraph. Consider for example the following system of equations:

$$\varrho D_t u_i + Q_i u_i = [\lambda(i) u_i + f_i(x,t,u_i) + g_i(x,t,u_1,u_2) - h_i(x,t)]\varrho \qquad (2.11)$$

for i=1,2 where now both $\lambda(1) \in \mathbf{R}$ and $\lambda(2) \in \mathbf{R}$ and Q_1 and Q_2 are both #-#$\tilde{H}$-related to L. The results in [LgS] for a system like that in (2.11) were for $\lambda(i) \leq \lambda_1^*$, and we did not know how to proceed for $\lambda(i) > \lambda_1^*$. Theorem 2 shows how to proceed and covers the existence of solutions u_1 and u_2 in $\tilde{H}$ for *all* $\lambda(1) \in \mathbf{R}$ and *all* $\lambda(2) \in \mathbf{R}$ for the system in (2.11).

Reaction-diffusion equations are important in mathematical biology when the diffusion coefficients μ_i in the system (2.6) are not equal. They give rise to what is then called the Turing mechanism in the theory of pattern formation often referred to as morphogenesis. (See the interesting article [Mu1] and a further discussion in [Mu2, pp. 375-379].)

2.3 Examples

It is our intention in this section to give six different examples of an L and a Q which qualify for Theorems 1 and 2 of this chapter. We will close this section with a reaction-diffusion system which is of activator-inhibitor type, an important concept in mathematical biology (e.g., see [Be, p.116], [EK, pp.295-9], or [Fi, pp.140-1]), and which is covered by Theorem 2.

As a first example of an L and a Q which qualify for said theorems, we work in dimension N$\geq$ 1 and take $\Omega \subset \mathbf{R}^N$ to be a bounded open connected set and $\Gamma = \partial\Omega$. Also, we take $\varrho = q = p_i = 1$ for $i = 1, ..., N$, and

$$Lu = -\sum_{i=1}^{N} D_i^2 u + u.$$

Then it is easy to see from [*GT*, *p.* 214] that (Ω, Γ) is a V_L−region.

We set

$$A_i\left(x,s,\xi\right) = \xi_i + \xi_i / \left(1 + s^2 + |\xi|^2\right)^{\frac{1}{2}} \quad \text{for i=1,...,N, and}$$

$$B_0\left(x,s,\xi\right) = 1 + 1/\left(1 + s^2 + |\xi|^2\right)^{\frac{1}{2}}$$

where $\xi = (\xi_1, ..., \xi_N)$ and $|\xi|^2 = \xi_1^2 + \cdots + \xi_N^2$. Also, we define

$$Qu = -\sum_{i=1}^{N} D_i A_i\left(x,u,Du\right) + B_0\left(x,u,Du\right)u.$$

It is clear that Q meets (Q-1), (Q-2), and (Q-3) of Chapter 1 and (Q-5) of Chapter 2.

Since

$$\partial[\frac{|\xi|^2}{2} + \left(1 + s^2 + |\xi|^2\right)^{\frac{1}{2}}]/\partial\xi_i = A_i\,(x, s, \xi)$$

for i=1,...,N, it follows from [KS, p. 16] that Q also meets (Q-4) of Chapter 1. We shall show that Q meets (Q-6) of Chapter 2, and that Q is both $\#\tilde{H}$-related to L (see (1.10)) and $\#$-$\#\tilde{H}$-related to L (see(2.3)). So Q will qualify for Theorem 1 and also Theorem 2 as long as the resonance part of the system is not present, i.e., the part of the system from i=1,...,N_1 is absent.

To see that Q meets (Q-6), we observe that if both $v, D_t v \in \tilde{C}^1_{p,\varrho}(\Omega, \Gamma)$, then

$$\mathcal{Q}^{\tilde{}}\,(v, D_t v) = \langle v, D_t v\rangle^{\tilde{}}_{\varrho} + \sum_{i=1}^{N} \langle D_i v, D_t D_i v\rangle^{\tilde{}}_{\varrho}$$

$$+ \int_{\Omega}\int_0^{2\pi} D_t[1 + |v|^2 + |D_1 v|^2 + \cdots + |D_N v|^2]^{\frac{1}{2}}.$$

It is follows from the periodicity of v and its first derivatives in t (i.e., v(x,t)=v(x,t+2π)) that each of the terms on the right-hand side of this last equation is zero. Therefore, $\mathcal{Q}^{\tilde{}}\,(v, D_t v) = 0$, and (Q-6) is established.

To show that Q is $\#\tilde{H}$-related to L , we observe with

$$|Du|^2 = \sum_{i=1}^{N} |D_i u|^2$$

that

$$\left|\mathcal{Q}^{\tilde{}}\,(u, v) - \mathcal{L}^{\tilde{}}\,(u, v)\right|$$

$$= \left|\int_{\tilde{\Omega}} \left[uv + \sum_{i=1}^{N} D_i u D_i v\right]\right| / \left[1 + u^2 + |Du|^2\right]^{\frac{1}{2}}$$

$$\leq \{\int_{\tilde{\Omega}} \frac{u^2 + |Du|^2}{1 + u^2 + |Du|^2}\}^{\frac{1}{2}} \left|\mathcal{L}^{\tilde{}}\,(v, v)\right|^{\frac{1}{2}}$$

$$\leq [2\pi\ meas\ \Omega]^{\frac{1}{2}}\ \left|\mathcal{L}^{\tilde{}}\,(v, v)\right|^{\frac{1}{2}}.$$

So (1.10)(i) follows immediately from (1.2) and this last computation. Similarly (1.10)(ii) follows. Hence, Q is indeed $\#\tilde{H}$-related to L. Also, from this last inequality (2.3)(ii) follows. So Q is also $\#$-$\#\tilde{H}$-related to L, and our remarks concerning this example are complete.

For our next example, we take $\Omega = \mathbf{R}^2$, Γ =empty set, q=0, $p_1(x) = p_2(x) = \varrho(x) = e^{-(x_1^2+x_2^2)}$. Then

$$Lu = -\sum_{i=1}^{2} D_i\left[e^{-(x_1^2+x_2^2)} D_i u\right],$$

and as we have shown in (1,3.2), (Ω,Γ) is a V_L−region, where $\varphi_1(x)$, the first eigenvalue of L, is a positive constant, and $\lambda_1 = 0$. We define $A_i\left(x, s, \xi\right)$ by (1,3.25).Thus

$$Qu = -\sum_{i=1}^{2} D_i e^{-(x_1^2+x_2^2)}\{D_i u + \frac{D_i u}{\left[1+|Du|^2\right]^{\frac{1}{2}}}\}, \tag{3.1}$$

and it follows from our observations in Chapter 1 that Q meets (Q-1)-(Q-5). This example is a slight modification of our previous one, and it is easy to see that

$$\mathcal{Q}^{\tilde{}}\,(v, D_t v) = \sum_{i=1}^{2} \langle D_i v, D_t D_i v\rangle_{\varrho}^{\tilde{}} + \int_{\Omega} \varrho \int_0^{2\pi} D_t[1+|Dv|^2]^{\frac{1}{2}}$$

for $v, D_t v \in \tilde{C}^1_{p,\varrho}(\Omega,\Gamma)$. The periodicity of v and its derivatives in t once again implies that the right-hand side of this last equation is zero. Therefore, $\mathcal{Q}^{\tilde{}}\,(v, D_t v) = 0$, and hence (Q-6) holds. Also, we see from (1,3.26) and (1,3.27) that

$$\left|\mathcal{Q}^{\tilde{}}\,(u,v) - \mathcal{L}^{\tilde{}}\,(u,v)\right| \le 2^{\frac{1}{2}}\pi \left|\mathcal{L}^{\tilde{}}\,(v,v)\right|^{\frac{1}{2}}.$$

Consequently, we see from(1.2) and this last inequality that Q is both $\#\tilde{H}$-related to L and $\#$-$\#\tilde{H}$-related to L. Therefore Q qualifies for Theorem 1. We now show that it partially qualifies for Theorem 2.

From (3.1), we see that because φ_1 is a positive constant

$$\mathcal{Q}^{\tilde{}}\,(v, \varphi_1) = 0 \;\; \forall v \in \tilde{H}.$$

But $\lambda_1 = 0$. Therefore,

$$\mathcal{Q}^{\tilde{}}\,(v, \varphi_1) = \lambda_1 \langle v, \varphi_1\rangle_{\varrho}^{\tilde{}} \;\; \forall v \in \tilde{H}.$$

Consequently, φ_1 is an $\tilde{\text{H}}$-λ_1- eigenfunction of Q. Since λ_1 is a simple eigenvalue of L, λ_1 is an $\tilde{\text{H}}$-L-pseudo-eigenvalue of Q (see(2.5)). Therefore this example also qualifies for Theorem 2 provided all the eigenvalues in the first part of the system (i.e., for i=1,...,N_1) are equal to zero, that is $\lambda_{j_o(i)} = \lambda_1 = 0$ for i=1,...,N_1.

Our next example will be of an L and a Q which qualify for Theorem 2 with every eigenvalue of L an $\tilde{H}$-L-pseudo-eigenvalue for Q. In particular, any number of the form

$$\left(m_1^2 + m_2^2\right)\pi^2 + 2n$$

where m_1 and m_2 are positive integers and n is a nonnegative integer will qualify as an $\tilde{H}$-L-pseudo-eigenvalue for Q.

To accomplish this, we use the example given in (1,3.3) and (1,3.30) and choose $\Omega \subset \mathbf{R}^3$ where

$$\Omega = \check{\Omega} \times \mathbf{R} \quad \text{and} \quad \check{\Omega} = (0,1)^2,$$

i.e., $\check{\Omega}$ is the square

$$\check{\Omega} = \{x : \ 0 < x_i < 1, \ i = 1, 2\}.$$

Also, we take $\Gamma = \partial\check{\Omega} \times \mathbf{R}$, $q = 0$, and $p_1(x) = p_2(x) = p_3(x) = \varrho(x) = e^{-x_3^2}$. Then

$$Lu = -\sum_{i=1}^{3} D_i\left[e^{-x_3^2} D_i u\right],$$

and with m=(m_1,m_2), all the eigenfunctions of L are of the form

$$\Phi_{mn}(x) = 2\sin m_1\pi x_1 \sin m_2\pi x_2 \ H_n(x_3)/(2^{-n}n!\pi^{\frac{1}{2}})^{\frac{1}{2}}$$

for n=0,1,..., m_i=1,2,..., with i=1, 2, and where H_n is the n-th Hermite polynomial. In particular,

$$L\Phi_{mn}(x) = [(m_1^2 + m_2^2)\pi^2 + 2n]e^{-x_3^2}\Phi_{mn}(x)$$

and we see that (Ω, Γ) is indeed a V_L−region.

Next, we assume that the real-valued function F(r)$\in C^0([0,\infty))$ meets (1,3.29)(i) and (ii) and that

$$A_1(x,s,\xi) = p_1^{\frac{1}{2}}\xi_1, \ A_2(x,s,\xi) = p_2^{\frac{1}{2}}\xi_2$$

$$A_3(x,s,\xi) = 2^{-1}\left[\,1 + F(|\,\xi_3\,|)\,\right]\xi_3 p_3^{\frac{1}{2}}$$

where $p_1 = p_2 = p_3 = e^{-x_3^2} = \varrho$ and $q = 0$, so that

$$Qu = -\sum_{i=1}^{2} D_i \left[e^{-x_3^2} D_i u\right] - 2^{-1} D_3 e^{-x_3^2}[1 + F(|D_3 u|)] D_3 u.$$

It is then clear that Q meets (Q-1)-(Q-5), and from the definition of L that

$$\mathcal{Q}^{\tilde{}}(u,v) - \mathcal{L}^{\tilde{}}(u,v) = -2^{-1} \int_{\tilde{\Omega}} e^{-x_3^2} \left[1 - F\left(|D_3 u|\right)\right] D_3 u D_3 v \qquad (3.2)$$

$\forall u, v \in \tilde{H}$. In particular, if both $v, D_t v \in \tilde{C}^1_{p,\varrho}(\Omega, \Gamma)$, then

$$\mathcal{Q}^{\tilde{}}(v, D_t v) = 2^{-1} \int_{\Omega} e^{-x_3^2} \sum_{i=1}^{2} \int_0^{2\pi} D_t \left|D_i v\right|^2$$

$$+4^{-1} \int_{\Omega} e^{-x_3^2} \int_0^{2\pi} D_t \left|D_3 v\right|^2 + 4^{-1} \int_{\Omega} e^{-x_3^2} \int_0^{2\pi} D_t G\left(|D_3 v|^2\right)$$

where

$$G(s) = \int_0^s F\left(r^{\frac{1}{2}}\right) dr.$$

The 2π-periodicity of v and its first derivatives in t implies that the right-hand side of this last equation for $\mathcal{Q}^{\tilde{}}(v, D_t v)$ is zero. Hence, $\mathcal{Q}^{\tilde{}}(v, D_t v)$ itself is zero, and we see that Q satisfies (Q-6).

Next, we see from (3.2) and the fact that F(r) meets (1,3.29)(i) and (ii) that there is a constant c such that

$$\left|\mathcal{Q}^{\tilde{}}(u,v) - \mathcal{L}^{\tilde{}}(u,v)\right| \leq c \int_{\tilde{\Omega}} e^{-x_3^2} \left|D_3 v\right|$$

$$\leq c\left[\int_{\tilde{\Omega}} e^{-x_3^2}\right]^{\frac{1}{2}} \left[\mathcal{L}^{\tilde{}}(v,v)\right]^{\frac{1}{2}} \qquad (3.3)$$

for u,v $\in \tilde{H}$. It follows from this inequality that Q is both $\#\tilde{H}$-related to L and $\#$-$\#\tilde{H}$-related to L. Therefore Q qualifies for Theorem 1. We now show that it also qualifies for Theorem 2, i.e., every λ_{j_o}-eigenvalue of L is an $\tilde{H}$-L-pseudo-eigenvalue for Q.

To accomplish, this suppose that λ_{j_o} is an eigenvalue of L of multiplicity j_1 with $\tilde{\varphi}^c_{j_o 0}, \ldots, \tilde{\varphi}^c_{j_o + j_1 - 1\, 0}$ the corresponding orthonormal eigenfunctions and with $P^o_{j_o}(u)$ the projection of u onto the subspace of $\tilde{H}$ spanned by these orthonormal functions. Suppose also that $\{u_n\}_{n=1}^{\infty}$ is a sequence with $u_n \in S_n^{\ddagger}$, $\|u_n\|_{\tilde{L}^2_{\varrho}} + \|u_n\|_{\mathcal{L}^{\check{}}} \to \infty$, and

$$\lim_{n\to\infty} \frac{\left\|u_n - P^o_{j_o}(u_n)\right\|_{\tilde{L}^2_{\varrho}} + \left\|u_n - P^o_{j_o}(u_n)\right\|_{\mathcal{L}^{\check{}}}}{\|u_n\|_{\tilde{L}^2_{\varrho}} + \|u_n\|_{\mathcal{L}^{\check{}}}} = 0 \tag{3.4}$$

where $\|u_n\|_{\mathcal{L}^{\check{}}}$ is defined by (2.4), and $S_n^{\ddagger}$ by (4.23). What we have to show is that

$$\lim_{n\to\infty} \frac{\mathcal{Q}^{\check{}}(u_n, P^o_{j_o}u_n) - \mathcal{L}^{\check{}}(u_n, P^o_{j_o}u_n)}{\|u_n\|_{\tilde{L}^2_{\varrho}} + \|u_n\|_{\mathcal{L}^{\check{}}}} = 0. \tag{3.5}$$

In order to establish (3.5), we observe from (1,3.29)(i) and (ii) that $0 \leq F(r) \leq 1$ and given $\varepsilon > 0$, there is an r_o such that

$$|1 - F(r)|\, r \leq \varepsilon \quad \text{for } \mathrm{r} \geq \mathrm{r}_o.$$

Therefore, it it follows from (3.2) that

$$\left|\mathcal{Q}^{\check{}}(u_n, u_n) - \mathcal{L}^{\check{}}(u_n, u_n)\right| \leq 2^{-1}\mathrm{r}_o^2 \int_{\tilde{\Omega}} e^{-x_3^2} + \varepsilon \int_{\tilde{\Omega}} e^{-x_3^2}\, |D_3 u_n|$$

$$\leq 2^{-1}\mathrm{r}_o^2 \int_{\tilde{\Omega}} e^{-x_3^2} + \varepsilon\Big[\int_{\tilde{\Omega}} e^{-x_3^2}\Big]^{\frac{1}{2}} \|u_n\|_{\mathcal{L}^{\check{}}}$$

Dividing both sides of this last inequality by $\|u_n\|_{\tilde{L}^2_{\varrho}} + \|u_n\|_{\mathcal{L}^{\check{}}}$ and passing to the limit as $\mathrm{n} \to \infty$, we see that

$$\lim_{n\to\infty} \frac{\left|\mathcal{Q}^{\check{}}(u_n, u_n) - \mathcal{L}^{\check{}}(u_n, u_n)\right|}{\|u_n\|_{\tilde{L}^2_{\varrho}} + \|u_n\|_{\mathcal{L}^{\check{}}}} \leq \varepsilon\Big[\int_{\tilde{\Omega}} e^{-x_3^2}\Big]^{\frac{1}{2}},$$

and therefore that

$$\lim_{n\to\infty} \frac{\left|\mathcal{Q}^{\check{}}(u_n, u_n) - \mathcal{L}^{\check{}}(u_n, u_n)\right|}{\|u_n\|_{\tilde{L}^2_{\varrho}} + \|u_n\|_{\mathcal{L}^{\check{}}}} = 0. \tag{3.6}$$

Now

$$\begin{aligned}\mathcal{Q}^{\check{}}\left(u_n, P^o_{j_o}u_n\right) - \mathcal{L}^{\check{}}\left(u_n, P^o_{j_o}u_n\right) &= \mathcal{Q}^{\check{}}(u_n, u_n) - \mathcal{L}^{\check{}}(u_n, u_n)\\ &\quad + \mathcal{L}^{\check{}}(u_n, v_n) - \mathcal{Q}^{\check{}}(u_n, v_n)\end{aligned}$$

where $v_n = u_n$ -$P^o_{j_o}u_n$. Dividing this last equation by $\|u_n\|_{\tilde{L}^2_\varrho} + \|u_n\|_{\mathcal{L}^{\check{}}}$ and passing to the limit as n→ ∞, we see from (3.3), (3.4) and (3.6)

$$\lim_{n\to\infty} \frac{\left|\mathcal{Q}^{\check{}}\left(u_n, P^o_{j_o}u_n\right) - \mathcal{L}^{\check{}}\left(u_n, P^o_{j_o}u_n\right)\right|}{\|u_n\|_{\tilde{L}^2_\varrho} + \|u_n\|_{\mathcal{L}^{\check{}}}} = 0.$$

But this is exactly (3.5). Therefore every eigenvalue of L is an $\tilde{\text{H}}$-L -pseudo-eigenvalue for Q, and this example qualifies also for Theorem 2.

Our next example will once again have all the eigenvalues of L as $\tilde{\text{H}}$-L -pseudo- eigenvalues of Q, but in this case q≠0 and B_0(x,u,Du) will depend in a nonlinear fashion on u and one of its derivatives. We take $\Omega \subset \mathbf{R}^2$ to be the rectangle

$$\Omega = (-1,1) \times (0,1),$$

$\Gamma = \{(t,1) : -1 \le t \le 1\}$, $p_1 = \left(1 - x_1^2\right)x_2$, $p_2 = x_2$, $q = x_2$, and $\varrho = x_2$. Then

$$Lu = -D_1\left(1 - x_1^2\right)x_2 D_1 u - D_2 x_2 D_2 u + x_2 u.$$

It then follows from our previous remarks in §3 of Chapter 1 that the eigenfunctions of L corresponding to the weight ϱ are of the form

$$\Phi_{mn}(x_1, x_2) = P_m(x_1)\, J_0(k_n x_2)$$

where k_n is n-th positive root of J_0 (s) and where m=0,1,..., and n=1,2,... . In particular,

$$L\Phi_{mn}(x_1, x_2) = [m(m+1) + k_n^2 + 1]x_2 \Phi_{mn}(x_1, x_2).$$

When this example is compared with the one in (1,3.10), we see that (Ω, Γ) is indeed a V_L-region.

Next, we set

$$A_1(x,s,\xi) = p_1^{\frac{1}{2}}\xi_1 \text{ and } A_2(x,s,\xi) = p_2^{\frac{1}{2}}\xi_2 \frac{1 + F[(s^2 + \xi_2^2)^{\frac{1}{2}}]}{2}$$

where the real-valued function F(r)∈C^0([0,∞)) meets (1,3.29)(i) and (ii). Also, we set

$$B_0(x,s,\xi) = 1 - \frac{1 - F[(s^2 + \xi_2^2)^{\frac{1}{2}}]}{2},$$

and observe that $B_0(x,s,\xi) \ge a_0(x)/2$ because $a_0(x) = 1$. As a consequence of the above,

$$Qu = Lu + \frac{D_2\left\{1 - F[\left(u^2 + |D_2 u|^2\right)^{\frac{1}{2}}]\right\} x_2 D_2 u}{2}$$

$$-\frac{\left\{1-F[\left(u^2+|D_2u|^2\right)^{\frac{1}{2}}]\right\}x_2u}{2}.$$

It then follows that

$$\mathcal{Q}^{\tilde{}}(u,v)-\mathcal{L}^{\tilde{}}(u,v)=-\frac{\int_{\tilde{\Omega}}x_2\left[1-F[\left(u^2+|D_2u|^2\right)^{\frac{1}{2}}]\right]D_2uD_2v}{2}$$

$$-\frac{\int_{\tilde{\Omega}}x_2\left[1-F[\left(u^2+|D_2u|^2\right)^{\frac{1}{2}}]\right]uv}{2} \tag{3.7}$$

for u,v∈$\tilde{\text{H}}$. Now from the above, it is clear that Q satisfies (Q-1)-(Q-5) of this chapter. We observe from this last equation that if both $v, D_tv\in\tilde{C}^1_{p,\varrho}(\Omega,\Gamma)$, then

$$\mathcal{Q}^{\tilde{}}(v,D_tv)=\mathcal{L}^{\tilde{}}(v,D_tv)-4^{-1}\int_{\tilde{\Omega}}x_2[D_t(v^2+|D_2v|^2)]$$

$$+4^{-1}\int_{\Omega}x_2\int_0^{2\pi}D_tG\left(v^2+|D_2v|^2\right)$$

where

$$G(s)=\int_0^s F\left(r^{\frac{1}{2}}\right)dr.$$

The $2\pi-$periodicity of v and its first derivatives in t implies that the right-hand side of the equation for $\mathcal{Q}^{\tilde{}}(v,D_tv)$ is zero. Hence, $\mathcal{Q}^{\tilde{}}(v,D_tv)$ itself is zero, and we see that Q satisfies (Q-6).

Next, we see from the fact that F(r) meets (1,3.29)(i) and (ii) that there is a positive constant c such that

$$|\,1-F[(s^2+r^2)^{\frac{1}{2}}]\,||r|\le c \text{ and } |\,1-F[(s^2+r^2)^{\frac{1}{2}}]\,||s|\le c$$

$\forall r,s\in\mathbf{R}$. Consequently, we obtain from (3.7) that

$$\left|\mathcal{Q}^{\tilde{}}(u,v)-\mathcal{L}^{\tilde{}}(u,v)\right|\le 2^{-1}c\int_{\tilde{\Omega}}|x_2|\left(|v|+|D_2v|\right)$$

$$\le c[\int_{\tilde{\Omega}}|x_2|]^{\frac{1}{2}}[\mathcal{L}^{\tilde{}}(v,v)]^{\frac{1}{2}} \tag{3.8}$$

$\forall u,v\in\tilde{H}$. It follows from the fact that $\varrho=q$, (1.2), and this last inequality, that the limit in (1.11)(i) is met. Furthermore, from our earlier observation that $B_0(x,s,\xi)\ge a_0(x)/2$, it follows that the condition in (1.11)(ii) is met.

Hence, Q is $\#\tilde{H}$-related to L, and therefore qualifies for Theorem 1. It also follows from this last inequality that Q is $\#$-$\#\tilde{H}$-related to L. We next show that every λ_{j_o}–eigenvalue of L is an H̃-L-pseudo-eigenvalue for Q. Therefore this example will also qualify for Theorem 2.

To show that every λ_{j_o}–eigenvalue of L is indeed an H̃-L-pseudo-eigenvalue for Q, suppose that λ_{j_o} is an eigenvalue of L of multiplicity j_1 with

$$\tilde{\varphi}^c_{j_o 0}, \ldots, \tilde{\varphi}^c_{j_o+j_1-1\ 0}$$

the corresponding orthonormal eigenfunctions and with $P^o_{j_o}(u)$ the projection of u onto the subspace of $\tilde{H}$ spanned by these orthonormal functions. Suppose also that $\{u_n\}_{n=1}^\infty$ is a sequence with $u_n \in S^\ddagger_n$, $\|u_n\|_{\tilde{L}^2_\varrho} + \|u_n\|_{\mathcal{L}^\sim} \to \infty$, and

$$\lim_{n\to\infty} \frac{\left\|u_n - P^o_{j_o}(u_n)\right\|_{\tilde{L}^2_\varrho} + \left\|u_n - P^o_{j_o}(u_n)\right\|_{\mathcal{L}^\sim}}{\|u_n\|_{\tilde{L}^2_\varrho} + \|u_n\|_{\mathcal{L}^\sim}} = 0 \tag{3.9}$$

where $\|u_n\|_{\mathcal{L}^\sim}$ is defined by (2.4), and $S^\ddagger_n$ by (4.23). What we have to show is that

$$\lim_{n\to\infty} \frac{\mathcal{Q}^\sim(u_n, P^o_{j_o}u_n) - \mathcal{L}^\sim(u_n, P^o_{j_o}u_n)}{\|u_n\|_{\tilde{L}^2_\varrho} + \|u_n\|_{\mathcal{L}^\sim}} = 0. \tag{3.10}$$

In order to establish (3.10), we observe from (1,3.29)(i) and (ii) that $0 \le F(r) \le 1$ and given $\varepsilon > 0$, there is an $\eta > 0$ such that

$$\left|1 - F\left((s^2+r^2)^{\frac{1}{2}}\right)\right| |r| \le \varepsilon \quad \text{for } |r| \ge \eta, \qquad \text{and}$$

$$\left|1 - F\left((s^2+r^2)^{\frac{1}{2}}\right)\right| |s| \le \varepsilon \quad \text{for } |s| \ge \eta.$$

Therefore, it follows from (3.7) that

$$\left|\mathcal{Q}^\sim(u_n,u_n) - \mathcal{L}^\sim(u_n,u_n)\right| \le \eta^2 \int_{\tilde{\Omega}} |x_2| + \varepsilon \int_{\tilde{\Omega}} |x_2| \left(|u_n| + |D_2 u_n|\right)$$

$$\le \eta^2 \int_{\tilde{\Omega}} |x_2| + 2\varepsilon [\int_{\tilde{\Omega}} |x_2|]^{\frac{1}{2}} \left\|u_n\right\|_{\mathcal{L}^\sim}$$

Dividing both sides of this last inequality by $\|u_n\|_{\tilde{L}^2_\varrho} + \|u_n\|_{\mathcal{L}^\sim}$ and passing to the limit as $n \to \infty$, we see that

$$\lim_{n\to\infty} \frac{\left|\mathcal{Q}^\sim(u_n,u_n) - \mathcal{L}^\sim(u_n,u_n)\right|}{\|u_n\|_{\tilde{L}^2_\varrho} + \|u_n\|_{\mathcal{L}^\sim}} \le 2\varepsilon [\int_{\tilde{\Omega}} |x_2|]^{\frac{1}{2}},$$

and therefore that

$$\lim_{n\to\infty}\frac{\left|\mathcal{Q}^{\tilde{}}(u_n,u_n)-\mathcal{L}^{\tilde{}}(u_n,u_n)\right|}{\|u_n\|_{\tilde{L}^2_\varrho}+\|u_n\|_{\mathcal{L}^{\tilde{}}}}=0. \tag{3.11}$$

Now

$$\begin{aligned}\mathcal{Q}^{\tilde{}}\left(u_n,P^o_{j_o}u_n\right)-\mathcal{L}^{\tilde{}}\left(u_n,P^o_{j_o}u_n\right) &= \mathcal{Q}^{\tilde{}}(u_n,u_n)-\mathcal{L}^{\tilde{}}(u_n,u_n)\\ &\quad+\mathcal{L}^{\tilde{}}(u_n,v_n)-\mathcal{Q}^{\tilde{}}(u_n,v_n)\end{aligned}$$

where $v_n = u_n$ -$P^o_{j_o}u_n$. Dividing this last equation by $\|u_n\|_{\tilde{L}^2_\varrho}+\|u_n\|_{\mathcal{L}^{\tilde{}}}$ and passing to the limit as n$\to\infty$, we see from (3.8), (3.9) and (3.11) that

$$\lim_{n\to\infty}\frac{\left|\mathcal{Q}^{\tilde{}}\left(u_n,P^o_{j_o}u_n\right)-\mathcal{L}^{\tilde{}}\left(u_n,P^o_{j_o}u_n\right)\right|}{\|u_n\|_{\tilde{L}^2_\varrho}+\|u_n\|_{\mathcal{L}^{\tilde{}}}}=0.$$

But this is exactly (3.10). Therefore every eigenvalue of L is an $\tilde{\text{H}}$-L -pseudo-eigenvalue for Q, and this example qualifies also for Theorem 2.

All the examples of the L that we have considered in this section up to now have been of the more restricted form governed by the equation in (1,3.1). We now provide an example which does not fall into this category, but belongs to the more general situation covered by the equation in (1,1.6). The corresponding Q that we shall provide will qualify for both Theorems 1 and 2. In particular, every eigenvalue of L will be an $\tilde{\text{H}}$-L -pseudo-eigenvalue for Q.

For this example, we work in $\mathbf{R}^3$, take $\check{\Omega}\subset\mathbf{R}^2$ to be a bounded open connected set, $\Omega=\check{\Omega}\times(-1,1)$ and $\Gamma=\partial\check{\Omega}\times[-1,1]$. Also, we take $p_1 = p_2 = \varrho = 1$, $q = 0$, $p_3(x) = \left(1-x_3^2\right)$ and $a_{ij}\in C^\infty(\check{\Omega})$ and meeting, in addition, the conditions in (1,1.5) with $\check{\Omega}$ replacing Ω. Operating at first in $\check{\Omega}$, we set

$$L^\flat w(x_1,x_2) = -\sum_{i=1}^2 D_i\sum_{j=1}^2 a_{ij}(x_1,x_2)\,D_jw(x_1,x_2)\ ,$$

and obtain a CONS, $\{\psi_m(x_1,x_2)\}_{m=1}^\infty$, in $\text{L}^2\left(\check{\Omega}\right)$ which is also in $\text{W}_0^{1,2}\left(\check{\Omega}\right)\cap C^\infty\left(\check{\Omega}\right)$ (See [*GT*, *p*.214] .). We designate the corresponding sequence of eigenvalues by $\{\eta_m\}_{m=1}^\infty$. Next we define

$$Lu(x_1,x_2,x_3) = -L^\flat u(x_1,x_2,x_3) - D_3\left(1-x_3^2\right)D_3u(x_1,x_2,x_3)$$

.for u$\in C^1_{p,q,\varrho}(\Omega)$. Then it is clear that

$$\left\{\psi_m(x_1,x_2)\,P_n(x_3)\,(2n+1)^{\frac{1}{2}}/\sqrt{2}\right\}_{m=1,n=0}^{\infty,\infty}$$

is a CONS for L$\varrho^2(\Omega)$ where P_n is the n-th Legendre polynomial. Furthermore, we have L$\psi_m(x_1,x_2)\,P_n(x_3) = [\eta_m + n(n+1)]\,\psi_m P_n$, and therefore (Ω,Γ) is a $V_L - region$.

Next, we define Q by

$$Qu = Lu - 2^{-1}D_3\left(1-x_3^2\right)\left[F\left(|D_3u|\right)-1\right]D_3u$$

where F meets the conditions in (1,3.29)(i) and (ii). It then follows that

$$\mathcal{Q}^{\tilde{}}(u,v) - \mathcal{L}^{\tilde{}}(u,v) = 2^{-1}\int_{\tilde{\Omega}}\left(1-x_3^2\right)\left[F\left(|D_3u|\right)-1\right]D_3uD_3v,$$

and exactly the same arguments that worked for the previous two examples show (see (3.2) and (3.7)) that Q is both #$\tilde{\text{H}}$ -related to L and #-#$\tilde{\text{H}}$-related to L. It also follows from the same arguments that worked for the previous two examples that every eigenvalue of L is also an $\tilde{\text{H}}$- L-pseudo-eigenvalue of Q. So this example qualifies for both Theorem 1 and 2 and represents a situation where the L involved is governed by the more general situation given by equation in (1,1.6).

For our last two examples in this section, we shall actually solve two different reaction-diffusion systems of two equations governed by Theorem 2. The first system will have its first equation resonant at the eigenvalue λ_2 and its second equation nonresonant and between the eigenvalues λ_3 and λ_4. The second system that we shall solve will have its first equation resonant at λ_2 and its second equation nonresonant and below the eigenvalue λ_1. This latter system will also be of activator-inhibitor type (e.g., see [Be, p. 116], [EK, pp. 295-9], or [Fi, pp. 140-1]).

We work with $\Omega \subset \mathbf{R}^2$ the square given by

$$\Omega = (-1,1)\times(-1,1) \quad \text{and } \Gamma = \text{empty set.}$$

Also, we take $p_1 = 2^{-1}\left(1-x_1^2\right)$, $p_2 = 1-x_2^2$, $q=1$, and $\varrho = 1$. So that

$$Lu = -D_1 2^{-1}\left(1-x_1^2\right)D_1u - D_2\left(1-x_2^2\right)D_2u + u.$$

Consequently, the eigenfunctions of L are products of Legendre polynomials given by

$$\Phi_{mn}(x_1,x_2) = P_m(x_1)\,P_n(x_2) \quad \text{m,n=0,1,2,...} .$$

In particular, we have

$$L\Phi_{mn} = \left[2^{-1}m\,(m+1) + n\,(n+1) + 1\right]\Phi_{mn}.$$

It is then clear that (Ω, Γ) is a V_L–region with $\varphi_1(x_1, x_2) = 2^{-1}$, $\varphi_2(x_1, x_2) = x_1 3^{\frac{1}{2}} 2^{-1}$, $\varphi_3(x_1, x_2) = x_2 3^{\frac{1}{2}} 2^{-1}$, and $\varphi_4(x_1, x_2) = x_1 x_2 3 2^{-1}$. Also, the corresponding eigenvalues are $\lambda_1 = 1$, $\lambda_2 = 2$, $\lambda_3 = 3$, $\lambda_4 = 4$.

We define Q_1 and Q_2 as follows:

$$Q_1 u = Lu + 4^{-1} D_1\left(1 - x_1^2\right)\left[1 - F_1\left(|D_1 u|\right)\right] D_1 u,$$

$$Q_2 u = Lu + \frac{D_2\left\{1 - F_2[\left(u^2 + |D_2 u|^2\right)^{\frac{1}{2}}]\right\}\left(1 - x_2^2\right) D_2 u}{2} - \frac{\left\{1 - F_2[\left(u^2 + |D_2 u|^2\right)^{\frac{1}{2}}]\right\}\left(1 - x_2^2\right) u}{2}$$

where $F_1, F_2 \in C^\infty([0, \infty))$ and satisfy (1,3.29)(i) and (ii). Then it is clear from the arguments given in the three previous examples of this section that both Q_1 and Q_2 satisfy (Q-1)-(Q-6). Also it is clear that both Q_1 and Q_2 are #$\tilde{\text{H}}$ -related to L and #-#$\tilde{\text{H}}$-related to L. Furthermore, it follows from the same arguments that worked in the three previous examples that every eigenvalue of L is also an $\tilde{\text{H}}$-L -pseudo- eigenvalue for both Q_1 and Q_2. So both Q_1 and Q_2 qualify for Theorem 2 (as well as Theorem 1).

We now look at a specific situation involving Theorem 2, namely the following:

$$D_t u_1 + Q_1 u_1 = 2u_1 + f_1(x, t, u_1) + g_1(x, t, u_1, u_2) - h_1(x, t) \qquad (3.12)$$

$$D_t u_2 + Q_2 u_2 = 7u_2/2 + f_2(x, t, u_2) + g_2(x, t, u_1, u_2) - h_2(x, t)$$

where $x = (x_1, x_2)$,

$$f_1(x, t, u_1) = \frac{u_1}{\left(1 + u_1^2\right)^{\frac{1}{2}}}, \quad g_1(x, t, u_1, u_2) = -\frac{arctan\ u_2}{4\left(1 + u_1^2\right)},$$

$$f_2(x, t, u_2) = e^{-u_2^2}/2, \quad g_2(x, t, u_1, u_2) = \frac{tanh\ u_1}{1 + u_2^2},$$

and $h_1(x_1, x_2, t)$, $h_2(x_1, x_2, t) \in C^0\{[-1, 1] \times [-1, 1] \times [0, 2\pi]\}$ with h_2 arbitrary and h_1 meeting the additional condition

$$-4\pi < \int_0^{2\pi} \int_{-1}^{1} \int_{-1}^{1} h_1(x_1, x_2, t)\, x_1 < 4\pi, \qquad (3.13)$$

e.g., $h_1(x_1,x_2,t)=x_1 sin^2 t - tx_1^2 + t^2 x_1 x_2$.

To see that Theorem 2 does give us a solution to the system in (3.12) under the conditions that we have prescribed for h_1 and h_2, we observe that both f_1 and f_2 meet (f-9) and (f-10) and that both g_1 and g_2 meet (g-3)-(g-5). Also, the 7/2 on the right-hand side in the second equation in (3.12) corresponds to the nonresonant condition $\lambda_3 + \frac{1}{2}$ where $\frac{1}{2} < \lambda_4 - \lambda_3 = 1$ which is the condition in (2.7). So the second equation in the system (3.12) indeed meets all the requirements in the hypothesis of Theorem 2.

It remains to check the conditions for the first equation. In this case, we have that the 2 on the right -hand side of the first equation corresponds to the resonant condition λ_2. So what is required is that the condition in (2.8) be met when i=1 and w is any λ_2-eigenfunction of L.

Now all the λ_2-eigenfunctions of L are of the form cx_1 where c is a constant. Also,

$$f_1(x,t,s) = \frac{s}{(1+s^2)^{\frac{1}{2}}}.$$

Therefore, it follows from (f-11) that

$$f_{1+}(x,t) = 1 \text{ and } f_1^-(x,t) = -1.$$

Hence, (2.8) will follow if we can show that

$$\int_{\tilde{\Omega}} h_1(x,t)w < \int_{\tilde{\Omega}\cap[w>0]} w - \int_{\tilde{\Omega}\cap[w<0]} w$$

where $w = x_1$ or $w = -x_1$ and $\tilde{\Omega} = [-1,1] \times [-1,1] \times [0,2\pi]$. An easy computation shows that this is precisely the same as the conditions in (3.13). Therefore, Theorem 2 shows that under the prescribed conditions that we have given for h_1 and h_2, a solution to the reaction-diffusion system (3.12) exists.

We next change the system in (3.12) to the following system:

$$D_t u_1 + Q_1 u_1 = 2u_1 + f_1(x,t,u_1) \ + g_1(x,t,u_1,u_2) - h_1(x,t) \qquad (3.14)$$

$$D_t u_2 + Q_2 u_2 = -3u_2 + f_2(x,t,u_2) \ + g_2(x,t,u_1,u_2) - h_2(x,t)$$

where everything remains the same as before except the 7/2 on the right-hand side of the second equation has been changed to -3. We observe that -3 corresponds to the nonresonant situation $\lambda_1 - 4 = 1 - 4 = -3$. So the conditions in the hypothesis of Theorem 2 are met in this case also, and hence a solution to the above system in (3.14) exists.

Setting $G_1(x, t, u_1, u_2)$ =the right-hand side of the first equation in the reaction-diffusion system (3.14) and $G_2(x, t, u_1, u_2)$ =the right -hand side of the second equation in (3.14), we find after an easy computation that

$$\partial G_1/\partial u_1 > 0, \ \partial G_2/\partial u_1 > 0, \partial G_1/\partial u_2 < 0, \partial G_2/\partial u_2 < 0$$

for $(x,t) \in \tilde{\Omega}$ and $u_1, u_2 \in \mathbf{R}$. Therefore, the reaction-diffusion system in (3.14) qualifies as an autocatalytic system (in the first variable) which is also of activator-inhibitor type, ([Be, p. 116], [EK, pp. 295-9], or [Fi, pp. 140-1]), and our discussion of these final examples governed by Theorem 2 is complete.

2.4 Fundamental lemmas

We are given L defined by (1.3) where both (1,1.5)(i)-(iv) and conditions $(V_L - 1)$ and $(V_L - 2)$ defined in Chapter 1 hold. From (V_L -1) it follows that

$$\{\tilde{\varphi}^c_{n,k}\}^{\infty}_{n=1,}{}^{,\infty}_{k=0} \cup \{\tilde{\varphi}^s_{n,k}\}^{\infty}_{n=1,}{}^{,\infty}_{k=1} \text{ is a CONS for } \tilde{L}^2_{\varrho} = L^2_{\varrho}(\tilde{\Omega}) \tag{4.1}$$

where

$$\begin{aligned} (i)\ \tilde{\varphi}^c_{nk}(x,t) &= \varphi_n(x)/(2\pi)^{1/2} & k = 0, n = 1, 2, \ldots, \\ &= \varphi_n(x) \cos kt/\pi^{1/2} & n, k = 1, 2, \ldots; \\ (ii)\ \tilde{\varphi}^s_{nk}(x,t) &= \varphi_n(x) \sin kt/\ \pi^{1/2} & n, k = 1, 2, \ldots . \end{aligned} \tag{4.2}$$

We note that both $\tilde{\varphi}^c_{nk}$ and $\tilde{\varphi}^s_{nk}$ are in $\tilde{\mathrm{H}}$.

Next, we define the bilinear form

$$\mathcal{L}^{\tilde{}}_1(u,v) = \mathcal{L}^{\tilde{}}(u,v) + < u, v >_{\tilde{\varrho}} \qquad \forall v \in \tilde{H}. \tag{4.3}$$

where $\mathcal{L}^{\tilde{}}(u,v)$ is given by (1.5). It is clear that $\mathcal{L}^{\tilde{}}_1(\cdot,\cdot)$ constitutes an inner product on $\tilde{H}$. It is also clear from (1, 1.5),(1.2), (1.5) and (4.3) that

$$\exists \text{ positive constants } K_1 \text{ and } K_2 \text{ such that} \tag{4.4}$$

$$K_1 \|u\|^2_{\tilde{H}} \leq \mathcal{L}^{\tilde{}}_1(u,u) + \|D_t u\|^2_{\tilde{L}^2_{\varrho}} \leq K_2 \|u\|^2_{\tilde{H}} \qquad \forall u \in \tilde{H}.$$

For u$\in L^2_\varrho(\tilde{\Omega})\left(=\tilde{L}^2_\varrho\right)$, we set

$$(i)\ \hat{u}^c(n,k) =< u, \tilde{\varphi}^c_{nk} >_{\tilde{\varrho}} = \int_{\tilde{\Omega}} u\tilde{\varphi}^c_{nk}\varrho, \tag{4.5}$$

$$(ii)\ \hat{u}^s(n,k) =< u, \tilde{\varphi}^s_{nk} >_{\tilde{\varrho}},$$

and observe from $(\mathrm{V}_L - 2)$ in Chapter 1 and (1.5) that for $v \in \tilde{H}$

$$\tilde{\mathcal{L}}(v, \tilde{\varphi}^s_{nk}) = \int_T sin\ kt\mathcal{L}(v, \varphi_n)/\sqrt{\pi}$$

$$= \lambda_n < v, \tilde{\varphi}^s_{nk} >_{\tilde{\varrho}}$$

with a similar computation for $\tilde{\mathcal{L}}(v, \tilde{\varphi}^c_{nk})$. Consequently it follows from (4.3) that

$$\tilde{\mathcal{L}_1}(v, \tilde{\varphi}^s_{nk}) = (\lambda_n + 1)\, \hat{v}^s\, (n,k)\, , \tag{4.6}$$
$$\tilde{\mathcal{L}_1}(v, \tilde{\varphi}^c_{nk}) = (\lambda_n + 1)\, \hat{v}^c\, (n,k)\, .$$

Hence we have from (4.1) and (4.5) that

$$\{\tilde{\varphi}^c_{nk}/(\lambda_n+1)^{\frac{1}{2}}\}^{\infty,\infty}_{n=1,k=0} \cup \{\tilde{\varphi}^s_{nk}/(\lambda_n+1)^{\frac{1}{2}}\}^{\infty,\infty}_{n=1,k=1} \text{is a CONS for } \tilde{H}^\diamond \tag{4.7}$$

where $\tilde{H}^\diamond$ is the Hilbert space completion of $\tilde{H}$ with respect to the inner product $\tilde{\mathcal{L}_1}(\cdot,\cdot)$.

The first lemma we establish is the following:

Lemma 1. *With $\{\tilde{\varphi}^c_{nk}\}^{\infty,\infty}_{n=1,n=0} \cup \{\tilde{\varphi}^s_{nk}\}^{\infty,\infty}_{n=1,n=1}$ the CONS for $L^2_\varrho(\tilde{\Omega})$ defined by (4.2), set*

$$\tau_n\,(v) = \sum_{j=1}^{n} \hat{v}^c(j,0)\tilde{\varphi}^c_{jo} + \sum_{j=1}^{n}\sum_{k=1}^{n} [\hat{v}^c(j,k)\tilde{\varphi}^c_{jk} + \hat{v}^s(j,k)\tilde{\varphi}^s_{jk}]. \tag{4.8}$$

where $\hat{v}^c(n,k)$ and $\hat{v}^s(n,k)$ are defined by (4.5) (i) and (ii) respectively. Then

$$\lim_{n\to\infty} \|\tau_n\,(v) - v\|_{\tilde{H}} = 0 \quad \forall v \in \tilde{H}. \tag{4.9}$$

It is clear from the hypothesis of the lemma that if $w \in L^2_\varrho(\tilde{\Omega})\left(=\tilde{L}^2_\varrho\right)$, then $\|\tau_n\,(w) - w\|_{\tilde{L}^2_\varrho} \to 0$ as $n \to \infty$. Now since $v \in \tilde{H}$, $D_t v \in L^2_\varrho(\tilde{\Omega})$ by (1.2), and an easy computation shows that $\tau_n\,(D_t v) = D_t\tau_n v$. Hence, it follows that

$$\lim_{n\to\infty} \|D_t\ \tau_n v - D_t v\|_{\tilde{L}^2_\varrho} = 0. \tag{4.10}$$

From (4.6), we next see that

$$\hat{v}^c(j,k)\tilde{\varphi}^c_{jk} = \tilde{\mathcal{L}_1}(v, \tilde{\varphi}^c_{jk}/(\lambda_n+1)^{\frac{1}{2}})\ \ \tilde{\varphi}^c_{jk}/(\lambda_n+1)^{\frac{1}{2}}. \tag{4.11}$$

and similarly for $\hat{v}^s(j,k)\tilde{\varphi}^s_{jk}$. Therefore it follows from (4.7) and (4.8) that

$$\lim_{n\to\infty} \tilde{\mathcal{L}_1}(\tau_n(v) - v, \tau_n(v) - v) = 0 \qquad \forall v \in \tilde{H}. \tag{4.12}$$

This fact joined with (4.4) and (4.10) gives (4.9), and the proof of the lemma is complete.

Next, we establish the following:

Lemma 2. *(i) If $v \in \tilde{H}$, then*

$$\tilde{\mathcal{L}_1}(v,v) + \|D_t v\|^2_{\tilde{L}^2_\varrho} = \sum_{n=1}^{\infty} |\ \hat{v}^c(n,0)\ |^2 (\lambda_n + 1) \tag{4.13}$$

$$+ \sum_{n=1}^{\infty}\sum_{k=1}^{\infty} [|\hat{v}^c(n,k)|^2 + |\hat{v}^s(n,k)|^2](\lambda_n + 1 + k^2).$$

(ii) If $v \in L^2_\varrho(\tilde{\Omega}) \left(= \tilde{L}^2_\varrho\right)$ and the right-hand side of (4.13) is

finite, then $v \in \tilde{H}$.

As we observed in the proof of Lemma 1, $\tilde{\mathcal{L}_1}(\cdot,\cdot)$ is an inner product on $\tilde{H}$ and furthermore (4.12) holds. Hence,

$$\lim_{n\to\infty} \tilde{\mathcal{L}_1}(\tau_n(v), \tau_n(v)) = \tilde{\mathcal{L}_1}(v,v) \tag{4.14}$$

From (4.6)-(4.8), we see that

$$\mathcal{L}_1(\tau_n(v), \tau_n(v)) + \|D_t(\tau_n v)\|^2_{\tilde{L}^2_\varrho} = \sum_{j=1}^{n} |\ \hat{v}^c(j,0)\ |^2 (\lambda_j + 1)$$

$$+ \sum_{j=1}^{n}\sum_{k=1}^{n} [|\hat{v}^c(j,k)|^2 + |\hat{v}^s(j,k)|^2](\lambda_j + 1 + k^2).$$

The equality in (4.13)(i)then follows from well-known facts about double series, (4.10), and (4.14).

To prove (ii), we observe that $\tau_n(v) \in \tilde{H}$ for every n. Also, we see from (4.8) that if $m > n$, then

$$\mathcal{L}_1(\tau_m(v) - \tau_n(v), \tau_m(v) - \tau_n(v)) + \|D_t[\tau_m(v) - \tau_n(v)]\|^2_{\tilde{L}^2_\varrho} \tag{4.15}$$

$$\leq \sum_{j=n+1}^{m} |\hat{v}^c(j,0)|^2 (\lambda_j + 1) + \left(\sum_{j=n+1}^{m}\sum_{k=1}^{\infty} + \sum_{k=n+1}^{m}\sum_{j=1}^{\infty}\right) b_{jk}$$

where $b_{jk} = [|\hat{v}^c(j,k)|^2 + |\hat{v}^s(j,k)|^2](\lambda_j + 1 + k^2)$. Since both the single sum and the double sum on the right-hand side of (4.13) are finite, using (4.4), it follows from well-known aspects of double series and the inequality in (4.15) that $\{\tau_n(v)\}_{n=1}^{\infty}$ is a Cauchy sequence with respect to the inner product $<\cdot,\cdot>_{\tilde{H}}$. Hence, there exists a $w \in \tilde{H}$ such that $\|\tau_n(v) - w\|_{\tilde{H}} \to 0$ as $n \to \infty$. Consequently, from (1.2) we see that $\|\tau_n(v) - w\|_{\tilde{L}^2_\varrho} \to 0$ as $n \to \infty$. But $v \in \tilde{L}^2_\varrho$. Therefore from (4.1), $\|\tau_n(v) - v\|_{\tilde{L}^2_\varrho} \to 0$. Hence, v=w, and v is indeed in $\tilde{H}$, which proves Lemma 2.

Next, we establish the following.

Lemma 3. *Let $\tilde{\Omega}, \varrho, p, q,$ and L be as in the hypothesis of Theorems 1 and 2 and assume that* (Ω, Γ) is a V_L–region. *Then $\tilde{H}$ is compactly imbedded in $L^2_\varrho\left(\tilde{\Omega}\right)$.*

It follows from (4.4) that the norm $[\mathcal{L}_1(u,u) + \|D_t u\|^2_{\tilde{L}^2_\varrho}]^{\frac{1}{2}}$ is equivalent on $\tilde{H}$ to the norm $\|u\|_{\tilde{H}}$. Therefore, if we are given a sequence $\{v_n\}_{n=1}^{\infty} \subset \tilde{H}$ with $\|v_n\|_{\tilde{H}} \leq K \;\; \forall n$, it follows that there exists a subsequence $\{v_{n_m}\}_{m=1}^{\infty}$ and a $v \in \tilde{H}$ such that:

$$(i)\ \lim_{m\to\infty} < v_{n_m}, w >_{\tilde{H}} = < v, w >_{\tilde{H}} \quad \forall w \in \tilde{H}; \tag{4.16}$$

$$(ii)\ \lim_{m\to\infty} \mathcal{L}_1(v_{n_m}, w) = \mathcal{L}_1(v, w) \qquad \forall w \in \tilde{H};$$

$$(iii)\ \ \mathcal{L}_1(v_{n_m}, v_{n_m}) + < D_t v_{n_m}, D_t v_{n_m} >_{\tilde{\varrho}} \leq K_1^2 \qquad \forall m;$$

$$(iv)\ \ \mathcal{L}_1(v, v) + < D_t v, D_t v >_{\tilde{\varrho}} \leq K_1^2;$$

$$(v)\ \lim_{m\to\infty} \hat{v}^c_{n_m}(j,k) = \hat{v}^c(j,k),\ \lim_{m\to\infty} \hat{v}^s_{n_m}(j,k) = \hat{v}^s(j,k) \ \ \forall (j,k)$$

where K_1 is a positive constant. It follows from (4.16) (iii) and (iv) that

$$\mathcal{L}_1(v_{n_m} - v, v_{n_m} - v) + \|\,(D_t v_{n_m} - v)\,\|^2_{\tilde{L}^2_\varrho} \leq 4K_1^2 \quad \forall m. \tag{4.17}$$

To complete the proof of the lemma, it suffices to show that

$$\lim_{m\to\infty} \|v_{n_m} - v\|^2_{\tilde{L}^2_\varrho} = 0 \tag{4.18}$$

To establish (4.18), we observe from (4.1) that

$$\|v_{n_m} - v\|^2_{\tilde{L}^2_\varrho} = \sum_{j=1}^{\infty} a_m(j) + \sum_{j=1}^{\infty}\sum_{k=1}^{\infty} b_m(j,k) \tag{4.19}$$

where

$$(i)\quad a_m(j) = |\ \hat{v}^c_{n_m}(j,0) - \hat{v}^c(j,0)\ |^2 \quad \text{for } j = 1, 2, \dots , \tag{4.20}$$

$$(ii)\quad b_m(j,k) = \left|\hat{v}^c_{n_m}(j,k) - \hat{v}^c(j,k)\right|^2 + \left|\hat{v}^s_{n_m}(j,k) - \hat{v}^s(j,k)\right|^2 \quad \text{for } j = 1, 2, \dots .$$

Now, we see from (4.13), (4.17), and (4.20)(i) that

$$\sum_{j=1}^{\infty} a_m(j)(\lambda_j + 1) \le 4K_1^2 \tag{4.21}$$

where $\{\lambda_j\}_{j=1}^{\infty}$ are given in $(V_L - 2)$ of Chapter I. Since $\lambda_j \le \lambda_{j+1}$, we see from (4.21) that

$$\sum_{j=1}^{\infty} a_m(j) \le \sum_{j=1}^{J} a_m(j) + \lambda_J^{-1} \sum_{j=J+1}^{\infty} a_m(j)\lambda_j$$

$$\le \sum_{j=1}^{J} a_m(j) + 4K_1^2/\lambda_J.$$

It therefore follows from (4.16)(v), (4.20)(i), and this last computation that

$$lim\ sup_{m\to\infty} \sum_{j=1}^{\infty} a_m(j) \le 4K_1^2/\lambda_J.$$

Since $\lambda_J \to \infty$ as $J \to \infty$, we conclude that

$$\lim_{m\to\infty} \sum_{j=1}^{\infty} a_m(j) = 0. \tag{4.22}$$

Next from (4.13), (4.17), and (4.20)(ii), we see that

$$\sum_{j=1}^{\infty}\sum_{k=1}^{\infty} b_m(j,k)\left(\lambda_j + 1 + k^2\right) \le 4K_1^2.$$

Therefore,

$$\sum_{j=1}^{\infty}\sum_{k=1}^{\infty} b_m(j,k) \le \sum_{j=1}^{J}\sum_{k=1}^{J} b_m(j,k) + \lambda_J^{-1} \sum_{j=J+1}^{\infty}\sum_{k=1}^{\infty} b_m(j,k)\lambda_J$$

$$+J^{-2} \sum_{k=J+1}^{\infty}\sum_{k=1}^{\infty} b_m(j,k)k^2.$$

$$\le \sum_{j=1}^{J}\sum_{k=1}^{J} b_m(j,k) + 4K_1^2\left[\lambda_J^{-1} + J^{-2}\right].$$

It follows, therefore, from (4.16)(v), (4.20)(ii), and this last computation that

$$lim\ sup_{m\to\infty} \sum_{j=1}^{\infty}\sum_{k=1}^{\infty} b_m(j,k) \le 4K_1^2\left[\lambda_J^{-1} + J^{-2}\right].$$

Since the right hand side of this last inequality goes to zero as $J \to \infty$, we conclude that

$$\lim_{m\to\infty} \sum_{j=1}^{\infty}\sum_{k=1}^{\infty} b_m(j,k) = 0$$

This last fact joined with (4.19) and (4.22) gives that

$$\lim_{m\to\infty} \|v_{n_m} - v\|^2_{\tilde{L}^2_\varrho} = 0.$$

Hence (4.18) is established and the proof of the lemma is complete.

Next, we define

$$S_n^\ddagger = \{v \in \tilde{H} : v = \sum_{j=1}^{n} \eta_{j0}^c \tilde{\varphi}_{j0}^c + \sum_{j=1}^{n}\sum_{k=1}^{n} \eta_{jk}^c \tilde{\varphi}_{jk}^c + \eta_{jk}^s \tilde{\varphi}_{jk}^s \tag{4.23}$$

$$\eta_{jk}^c, \eta_{jk}^s \in \mathbf{R},\ j = 1, \cdots, n, k = 0, \cdots, n\} \ .$$

Using this subspace, we then prove the following remark:

Remark 2 *If* $u_n \in S_n^\ddagger$, *then* $\mathcal{Q}^{\tilde{}}(u_n, D_t u_n) = 0$.

To establish the above remark, we observe that if $u_n \in S_n^\ddagger$, then $D_t u_n \in S_n^\ddagger$. Also it is clear that there exists a sequence $\{v_k\}_{k=1}^\infty$ with $v_k \in \tilde{C}_{p,\varrho}^{1b}$ such that

$$\lim_{k\to\infty} [\|v_k - u_n\|_{\tilde{H}} + \|D_t v_k - D_t u_n\|_{\tilde{H}}] = 0.$$

This last limit along with (Q-1), (Q-2), (1.1), and (1.9) imply that there exists a subsequence $\{v_{k_j}\}_{j=1}^\infty$ such that

$$\lim_{j\to\infty} \mathcal{Q}^{\tilde{}}(v_{k_j}, D_t v_{k_j}) = \mathcal{Q}^{\tilde{}}(u_n, D_t u_n).$$

But by (Q-6), $\mathcal{Q}^{\tilde{}}(v_{k_j}, D_t v_{k_j}) = 0 \ \forall j$, and the above remark is established.

Using the subspace defined in (4.23) above, we next prove the following lemma:

Lemma 4. *Assume that all the conditions in the hypothesis of Theorem 1 hold except possibly for (1, 1.23) and (1, 1.24). Let* $S_n^\ddagger$ *be the subspace of* $\tilde{H}$ *defined by (4.23). Take* $n_o = j_0 + j_1$. *Then for every* $n \geq n_o$, *there is a* $u_n \in S_n^\ddagger$ *with the property that*

$$< D_t u_n, v >_{\varrho}^{\tilde{}} + \mathcal{Q}^{\tilde{}}(u_n, v) = (\lambda_{j_o} + \gamma n^{-1}) < u_n, v >_{\varrho}^{\tilde{}} \tag{4.24}$$

$$+(1 - n^{-1}) < f(\cdot, u_n) + g(\cdot, \cdot, u_n), v >_{\varrho}^{\tilde{}} \quad \forall v \in S_n^\ddagger.$$

To prove the lemma, we first observe that $D_t \tilde{\varphi}_{n0}^c = 0$ and that

$$D_t \tilde{\varphi}_{nk}^c = -k \tilde{\varphi}_{nk}^s, \quad D_t \tilde{\varphi}_{nk}^s = k \tilde{\varphi}_{nk}^c \quad for\ k \geq 1.$$

Hence, it follows from (4.23) that

$$(i) \quad v \in S_n^\ddagger \Longrightarrow D_t v \in S_n^\ddagger, \tag{4.25}$$

$$(ii) < D_t(\alpha\tilde{\varphi}_{jk}^c + \beta\tilde{\varphi}_{jk}^s), \alpha\tilde{\varphi}_{jk}^c + \beta\tilde{\varphi}_{jk}^s >_{\varrho}^{\tilde{}}$$

$$= 0 \ \text{ for } j, k \geq 1 \ \text{ and } \alpha, \beta \in \mathbf{R}.$$

Next, we let $\{\psi_i\}_{i=1}^{2n^2+n}$ be an enumeration of

$$\{\tilde{\varphi}_{jk}^c\}_{j=1,k=0}^{n,\ \ n} \cup \{\tilde{\varphi}_{jk}^s\}_{j=1,k=1}^{n,\ \ n}$$

where $n \geq j_o + j_1$. Also we set

$$n^\ddagger = (j_o + j_1 - 1)(2n + 1) \tag{4.26}$$

and in the above enumeration $\{\psi_i\}_{i=1}^{n^\ddagger}$ is understood to be an enumeration of

$$\{\tilde{\varphi}^c_{jk}\}_{j=1,\quad k=0}^{j_o+j_1-1,\ n} \cup \{\tilde{\varphi}^s_{jk}\}_{j=1,\quad k=1}^{j_o+j_1-1,\ n}.$$

With this enumeration defined, we set

$$u = \sum_{i=1}^{2n^2+n} \alpha_i \psi_i \text{ and } u^{\approx} = \sum_{i=1}^{2n^2+n} \delta_i \alpha_i \psi_i \tag{4.27}$$

where

$$\delta_i = \{^{-1 \quad \text{for } 1 \le i \le n^\ddagger}_{1 \quad \text{for } n^\ddagger + 1 \le i \le 2n^2+n} \quad , \tag{4.28}$$

and define

$$F_i(\alpha) = < D_t u, \delta_i \psi_i >_{\tilde{\varrho}} + \mathcal{Q}^{\tilde{}}(u, \delta_i \psi_i) - (\lambda_{j_o} + \gamma n^{-1}) < u, \delta_i \psi_i >_{\tilde{\varrho}} \tag{4.29}$$

$$-(1 - n^{-1}) < f(\cdot, u) + g(\cdot, \cdot, u), \delta_i \psi_i >_{\tilde{\varrho}}$$

where $\alpha = (\alpha_1, \cdots, \alpha_{2n^2+n})$ and $\gamma = (\lambda_{j_o+j_1} - \lambda_{j_o})/2$.

Now with u and $u^{\approx}$ defined in (4.27), we see that $u = u_1 + u_2$ and $u^{\approx} = -u_1 + u_2$ where $u_1 = \sum_{i=1}^{n^\ddagger} \alpha_i \psi_i$ and $n^\ddagger$ is defined in (4.26). It is clear from orthogonality that $< D_t u_1, u_2 >_{\tilde{\varrho}} = 0$ and $< D_t u_2, u_1 >_{\tilde{\varrho}} = 0$. Also, it is clear from (4.25)(ii) that $< D_t u_1, u_1 >_{\tilde{\varrho}} = 0$ and $< D_t u_2, u_2 >_{\tilde{\varrho}} = 0$. Hence in particular, it follows that $< D_t u, u^{\approx} >_{\tilde{\varrho}} = 0$ and consequently from (4.27) and (4.29) that

$$\sum_{i=1}^{2n^2+n} F_i(\alpha)\alpha_i = \mathcal{Q}^{\tilde{}}(u, u^{\approx}) - (\lambda_{j_o} + \gamma n^{-1}) < u, u^{\approx} >_{\tilde{\varrho}} \tag{4.30}$$

$$-(1 - n^{-1}) < f(\cdot, u) + g(\cdot, \cdot, u), u^{\approx} >_{\tilde{\varrho}}.$$

Now we see from (4.26) -(4.28)

$$u^{\approx} = \sum_{j=1}^{n} \varsigma_j \hat{u}^c(j,0)\tilde{\varphi}^c_{j0} + \sum_{j=1}^{n}\sum_{k=1}^{n} \varsigma_j \left[\hat{u}^c(j,k)\tilde{\varphi}^c_{jk} + \hat{u}^s(j,k)\tilde{\varphi}^s_{jk}\right], \tag{4.31}$$

$$u = \sum_{i=1}^{n} \hat{u}^c(j,0)\tilde{\varphi}^c_{j0} + \sum_{j=1}^{n}\sum_{k=1}^{n} \hat{u}^c(j,k)\tilde{\varphi}^c_{jk} + \hat{u}^s(j,k)\tilde{\varphi}^s_{jk},$$

$$\varsigma_j = \{^{-1 \quad 1 \le j \le j_o+j_1-1}_{1 \quad j_o+j_1 \le j \le n}.$$

Consequently, we obtain from (4.1), (4.3), (4.6) and the fact that $\mathcal{L}^{\tilde{}}(\cdot,\cdot)$ is linear in both variables that

$$\mathcal{L}^{\tilde{}}(u, u^{\approx}) = \sum_{j=1}^{n} \varsigma_j \lambda_j \left|\hat{u}^c(j,0)\right|^2 + \sum_{j=1}^{n}\sum_{k=1}^{n} \left[\left|\hat{u}^c(j,k)\right|^2 + \left|\hat{u}^s(j,k)\right|^2\right] \lambda_j \varsigma_j$$

Adding and subtracting $-\gamma < u, u^{\approx} >_{\varrho}^{\tilde{}} + \mathcal{L}^{\tilde{}}(u, u^{\approx})$ to the right-hand side of (4.30), we see from this last computation that

$$\sum_{i=1}^{2n^2+n} F_i(\alpha)\alpha_i = \sum_{j=1}^{n} \varsigma_j[\lambda_j - \lambda_{j_o} - \gamma] \left|\hat{u}^c(j,0)\right|^2 \tag{4.32}$$

$$+\sum_{j=1}^{n}\sum_{k=1}^{n} \varsigma_j[\lambda_j - \lambda_{j_o} - \gamma][\left|\hat{u}^c(j,k)\right|^2 + \left|\hat{u}^s(j,k)\right|^2]$$

$$-(1-n^{-1}) < f(\cdot,u) - \gamma u, u^{\approx} >_{\varrho}^{\tilde{}} -(1-n^{-1}) < g(\cdot,\cdot,u), u^{\approx} >_{\varrho}^{\tilde{}}$$

$$+\mathcal{Q}^{\tilde{}}(u, u^{\approx}) - \mathcal{L}^{\tilde{}}(u, u^{\approx}).$$

Observing from (4.31) that $\varsigma_j[\lambda_j - \lambda_{j_o} - \gamma] \geq \gamma$ for $j = 1, \cdots, n$ (because $\lambda_{j_o+j_1} - \gamma = \lambda_{j_o} + \gamma$) and from (f-2) in Chapter 1 and (4.1) that

$$\begin{aligned} \left|< f(\cdot,u) - \gamma u, u^{\approx} >_{\varrho}^{\tilde{}}\right| &\leq \gamma \left\|u\right\|_{\tilde{L}^2_{\varrho}} \left\|u^{\approx}\right\|_{\tilde{L}^2_{\varrho}} + K \left\|u^{\approx}\right\|_{\tilde{L}^2_{\varrho}} \\ &\leq \gamma \left\|u\right\|^2_{\tilde{L}^2_{\varrho}} + K \left\|u\right\|_{\tilde{L}^2_{\varrho}} \end{aligned}$$

where $K = [\int_T \int_\Omega |f_o|^2]^{\frac{1}{2}}$, we obtain from (4.32) that

$$\sum_{i=1}^{2n^2+n} F_i(\alpha)\alpha_i \geq \gamma n^{-1} \left\|u\right\|^2_{\tilde{L}^2_{\varrho}} - (1-n^{-1}) < g(\cdot,\cdot,u), u^{\approx} >_{\varrho}^{\tilde{}} \tag{4.33}$$

$$-K \left\|u\right\|_{\tilde{L}^2_{\varrho}} + \mathcal{Q}^{\tilde{}}(u, u^{\approx}) - \mathcal{L}^{\tilde{}}(u, u^{\approx}).$$

Now from (g-2), it follows that

$$\lim_{\left\|u\right\|_{\tilde{L}^2_{\varrho}} \to\infty} \left|< g(\cdot,\cdot,u), u^{\approx} >_{\varrho}^{\tilde{}}\right| \left\|u\right\|^{-2}_{\tilde{L}^2_{\varrho}} = 0 \tag{4.34}$$

Also, it follows from (4.31), (4.13), and (4.4) (since n is fixed) that

$$\exists K_1 > 0 \text{ s.t. } \left\|u^{\approx}\right\|_{\tilde{H}} \leq K_1 \left\|u\right\|_{\tilde{L}^2_{\varrho}} \qquad \forall u \in S_n^{\ddagger}.$$

Hence, it follows from (1.10)(i) that

$$\lim_{\|u\|_{\tilde{L}^2_\varrho}\to\infty} \left|\mathcal{Q}^{\sim}(u,u^{\approx})-\mathcal{L}^{\sim}(u,u^{\approx})\right| \|u\|^{-2}_{\tilde{L}^2_\varrho}=0 \qquad u\in S^{\ddagger}_n. \tag{4.35}$$

Also, we see from (4.27) that $\|u\|^2_{\tilde{L}^2_\varrho}=|\alpha|^2$. We conclude from (4.33)-(4.35) that there exists a positive constant s_o such that

$$\sum_{i=1}^{2n^2+n} F_i(\alpha)\alpha_i \geq \gamma n^{-1}|\alpha|^2 2^{-1} \qquad \text{for } |\alpha|>s_o. \tag{4.36}$$

Next, we observe from (Q-1)-(Q-5), (f-2), (g-1), (g-2), and (4.27)-(4.29) that

$$F_i:\mathbf{R}^{2n^2+n}\to\mathbf{R} \text{ is continuous in the uniform topology}$$

for $i=1,\ldots,2n^2+n$. We consequently obtain from (4.36) (See [Ke, p.219] or [Ni, p.18].) that there exists $\alpha^*=(\alpha^*_1,\cdots,\alpha^*_{2n^2+n})$ such that $F_i(\alpha^*)=0$ for $i=1,\cdots,2n^2+n$. In particular, $-F_i(\alpha^*)-0$ for $i=1,\cdots,n^\ddagger$ where $n^\ddagger$ is given (4.26) and $F_i(\alpha^*)=0$ for $i=n^\ddagger+1,\cdots,2n^2+n$. Taking $u_n=\sum_{i=1}^{2n^2+n}\alpha^*_i\psi_i$, we consequently obtain from (4.29) that

$$\begin{aligned}&< D_t u_n,\psi_i>^{\sim}_\varrho+\mathcal{Q}^{\sim}(u_n,\psi_i)=(\lambda_{j_o}+\gamma n^{-1}) < u_n,\psi_i>^{\sim}_\varrho\\&\qquad+(1-n^{-1}) < f(\cdot,u_n)+g(\cdot,\cdot,u_n),\psi_i>^{\sim}_\varrho.\end{aligned}$$

for $i=1,\cdots,2n^2+n$. Hence it follows from the definition of $S^{\ddagger}_n$ that (4.24) holds, and the proof of the lemma is complete.

Next, we establish a lemma which will be needed in the proof of Theorem 2. This lemma will deal with the following perturbation of the system given in (2.6):

$$\begin{aligned}\varrho D_t u_i+\mu_i Q_i u_i&=[\mu_i(\lambda_{j_o(i)}+\frac{1}{n})u_i+f_i(x,t,u_i)\\&\quad+g_i(x,t,u_1,\ldots,u_{N_1+N_2+N_3})-h_i\,]\,\varrho \qquad i=1,\ldots,N_1\\&=\left[\mu_i(\lambda_{j_o(i)}+\varepsilon_i)u_i+f_i(x,t,u_i)+g_i(x,t,u_1,\ldots,u_{N_1+N_2+N_3})-h_i\right]\varrho\\&\qquad i=N_1+1,\ldots,N_1+N_2\end{aligned}$$

$$= [\mu_i(\lambda_1 - \varepsilon_i)u_i + f_i(x,t,u_i) + g_i(x,t,u_1,\ldots,u_{N_1+N_2+N_3}) - h_i]\varrho$$

$$i = N_1 + N_2 + 1, \ldots, N_1 + N_2 + N_3.$$

However, we will deal with this system in the following weak solution form:

$$< D_t u_i, v >_{\varrho}^{\sim} + \mu_i \tilde{Q}_i(u_i,v) = \tag{4.37}$$

$$\mu_i(\lambda_{j_o(i)} + \frac{1}{n}) < u_i, v >_{\varrho}^{\sim} + < f_i(\cdot,\cdot,u_i) + g_i(\cdot,\cdot,u_1,\ldots,u_{N_1+N_2+N_3})$$

$$-h_i(\cdot,\cdot), v >_{\varrho}^{\sim} \quad \forall v \in S_n^{\ddagger}, \qquad i = 1,\ldots,N_1,$$

$$\mu_i(\lambda_{j_o(i)} + \varepsilon_i) < u_i, v >_{\varrho}^{\sim} + < f_i(\cdot,\cdot,u_i) + g_i(\cdot,\cdot,u_1,\ldots,u_{N_1+N_2+N_3})$$

$$-h_i(\cdot,\cdot), v >_{\varrho}^{\sim} \quad \forall v \in S_n^{\ddagger}, \ i = N_1 + 1,\ldots,N_1 + N_2,$$

$$\mu_i(\lambda_1 - \varepsilon_i) < u_i, v >_{\varrho}^{\sim} + < f_i(\cdot,\cdot,u_i) + g_i(\cdot,\cdot,u_1,\ldots,u_{N_1+N_2+N_3})$$

$$-h_i(\cdot,\cdot), v >_{\varrho}^{\sim} \quad \forall v \in S_n^{\ddagger}, \ i = N_1 + N_2 + 1,\ldots,N_1 + N_2 + N_3.$$

In order to state the lemma that we need, we shall set

$$\varepsilon_0^* = min\left(\gamma_1,\ldots,\gamma_{N_1},\varepsilon_{N_1+1},\ldots,\varepsilon_{N_1+N_2+N_3}\right), \tag{4.38}$$

$$\varepsilon_1^* = min\left(2\gamma_{N_1+1} - \varepsilon_{N_1+1},\ldots,2\gamma_{N_1+N_2} - \varepsilon_{N_1+N_2}\right)$$

where $\gamma_i = (\lambda_{j_o(i)+j_1(i)} - \lambda_{j_o(i)})/2$ for $i = 1,\ldots,N_1 + N_2$, and $\varepsilon_i > 0$ for $i = N_1+1,\ldots,N_1+N_2+N_3$ and ε_i meets (2.7) for $i = N_1+1,\ldots,N_1+N_2$. It is clear that both $\varepsilon_0^* > 0$ and $\varepsilon_1^* > 0$. Also, n_o^* will be an integer such that

$$n_o^* \geq max\left(j_0(i) + j_1(i), 1/\varepsilon_0^*, 1/\varepsilon_1^*\right) \quad i = 1,\ldots,N_1 + N_2. \tag{4.39}$$

The lemma that we shall establish will be the following:

Lemma 5. *Assume that all the conditions in the hypothesis of Theorem 2 hold except possibly for (2.8). Assume also that n_o^* is an integer that meets (4.39). Then for every $n > n_o^*$,*

$$\exists \left(u_{1n}^*,\ldots,u_{N_1+N_2+N_3 n}^*\right) \text{ with } u_{in}^* \in S_n^{\ddagger} \ i = 1,\ldots,N_1 + N_2 + N_3$$

such that (4.37) holds for every $v \in S_n^{\ddagger}$ where $S_n^{\ddagger}$ is the subspace of $\tilde{H}$ defined by (4.23).

To prove Lemma 5, we use the same notation that we set up for the proof of Lemma 4, but we want to be more specific about the enumeration of the ψ_i's given in the paragraph below (4.25). In particular, with n fixed, we want the enumeration of the ψ_i's to be in blocks of 2n+1 in the following manner:

$$\psi_1 = \tilde{\varphi}_{10}^c,\ \psi_2 = \tilde{\varphi}_{11}^c, \ldots,\ \psi_{n+1} = \tilde{\varphi}_{1n}^c,\ \psi_{n+2} = \tilde{\varphi}_{11}^s, \ldots,\ \psi_{2n+1} = \tilde{\varphi}_{1n}^s$$

$$\psi_{(2n+1)+1} = \tilde{\varphi}_{20}^c, \ldots,\ \psi_{(2n+1)+n+1} = \tilde{\varphi}_{2n}^c, \ldots,\ \psi_{(2n+1)2} = \tilde{\varphi}_{2n}^s, \text{ etc.}$$

The general formula that will cover this enumeration of the ψ_i's is

$$\psi_{(2n+1)(j-1)+k+1} = \tilde{\varphi}_{jk}^c \quad j = 1,\ldots,n, \quad k = 0,\ldots,n, \tag{4.40}$$

$$\psi_{(2n+1)(j-1)+k+1} = \tilde{\varphi}_{j(k-n)}^s \quad j = 1,\ldots,n, \quad k = n+1,\ldots,2n.$$

Also, we set

$$n^{\ddagger}(i) = [j_o(i) + j_1(i) - 1](2n+1) \tag{4.41}$$

for $i = 1, \ldots, N_1 + N_2$ and observe that $\{\psi_l\}_{l=1}^{n^{\ddagger}(i)}$ is an enumeration of

$$\{\tilde{\varphi}_{jk}^c\}_{j=1,\quad k=0}^{j_o(i)+j_1(i)-1,\ n} \cup \{\tilde{\varphi}_{jk}^s\}_{j=1,\quad k=1}^{j_o(i)+j_1(i)-1,\ n}.$$

Next, we define

$$\delta_k(i) = \{ {}^{-1 \quad \text{for } 1 \le k \le n^{\ddagger}(i)}_{1 \quad \text{for } n^{\ddagger}(i)+1 \le k \le 2n^2+n} \quad \text{for } i = 1, \ldots, N_1 + N_2, \tag{4.42}$$

$$\delta_k(i) = 1 \qquad \text{for } i = N_1 + N_2 + 1, \ldots, N_1 + N_2 + N_3,$$

and as in (4.27) in the proof of Lemma 4, set

$$u_{in} = \sum_{k=1}^{2n^2+n} \alpha_k^i \psi_k \text{ and } u_{in}^{\approx} = \sum_{k=1}^{2n^2+n} \delta_k(i)\, \alpha_k^i \psi_k. \tag{4.43}$$

Now with

$$\alpha = \left(\alpha_k^i\right)_{k=1,\ldots,2n^2+n, i=1,\ldots,N_1+N_2+N_3} \in \mathbf{R}^{(2n^2+n)(N_1+N_2+N_3)},$$

we define $B_k^i(\alpha)$ as follows:

$$B_k^i(\alpha) - < D_t u_{in}, \delta_k(i)\, \psi_k >_{\tilde{\varrho}} - \mu_i \tilde{\mathcal{Q}}_i(u_{in}, \delta_k(i)\, \psi_k) \tag{4.44}$$

$$= -\mu_i(\lambda_{j_o(i)} + \frac{1}{n}) < u_{in}, \delta_k(i)\, \psi_k >_{\tilde{\varrho}} - < f_i(\cdot,\cdot,u_{in}) +$$

$$g_i(\cdot,\cdot,u_{1n},\ldots,u_{N_1+N_2+N_3 n}) - h_i(\cdot,\cdot), \delta_k(i)\, \psi_k >_{\tilde{\varrho}} \qquad i = 1, \ldots, N_1,$$

$$= -\mu_i(\lambda_{j_o(i)} + \varepsilon_i) < u_{in}, \delta_k(i)\, \psi_k >_{\tilde{\varrho}} - < f_i(\cdot,\cdot,u_{in}) + g_i(\cdot,\cdot,u_{1n},$$

$$\ldots, u_{N_1+N_2+N_3n}) - h_i(\cdot,\cdot), \delta_k\,(i)\,\psi_k >_{\varrho}^{\tilde{}} \quad i = N_1+1, \ldots, N_1+N_2,$$

$$= -\mu_i(\lambda_1 - \varepsilon_i) < u_{in}, \delta_k\,(i)\,\psi_k >_{\varrho}^{\tilde{}} - < f_i\,(\cdot,\cdot, u_{in}) + g_i(\cdot,\cdot, u_{1n}, \ldots,$$

$$u_{N_1+N_2+N_3n}) - h_i(\cdot,\cdot), \delta_k\,(i)\,\psi_k >_{\varrho}^{\tilde{}} \quad i = N_1+N_2+1, \ldots, N_1+N_2+N_3.$$

for k=1,...,2n$^2+1$ where u_{in} is a function of α and is defined in(4.43) and $\delta_k\,(i)$ is defined in(4.42).

Set $B(\alpha) = (B_k^i\,(\alpha))_{k=1,\ldots,2n^2+n, i=1,\ldots,N_1+N_2+N_3}$. Observing that in the second variable $\mathcal{Q}_i^{\tilde{}}\,(\cdot,\cdot)$ is linear, we have

$$B(\alpha)\cdot\alpha = \sum_{i=1}^{N_1+N_2+N_3}\sum_{k=1}^{2n^2+n} B_k^i\,(\alpha)\,\alpha_k^i \tag{4.45}$$

$$= \sum_{i=1}^{N_1}[< D_t u_{in}, u_{in}^{\approx} >_{\varrho}^{\tilde{}} + \mu_i \mathcal{Q}_i^{\tilde{}}\,(u_{in}, u_{in}^{\approx}) - \mu_i(\lambda_{j_o(i)} + \frac{1}{n}) < u_{in}, u_{in}^{\approx} >_{\varrho}^{\tilde{}}$$

$$- < f_i\,(\cdot,\cdot, u_{in}) + g_i(\cdot,\cdot, u_{1n}, \ldots, u_{N_1+N_2+N_3n}) - h_i(\cdot,\cdot), u_{in}^{\approx} >_{\varrho}^{\tilde{}}]+$$

$$\sum_{i=N_1+1}^{N_1+N_2}[< D_t u_{in}, u_{in}^{\approx} >_{\varrho}^{\tilde{}} + \mu_i \mathcal{Q}_i^{\tilde{}}\,(u_{in}, u_{in}^{\approx}) - \mu_i(\lambda_{j_o(i)} + \varepsilon_i) < u_{in}, u_{in}^{\approx} >_{\varrho}^{\tilde{}}$$

$$- < f_i\,(\cdot,\cdot, u_{in}) + g_i(\cdot,\cdot, u_{1n}, \ldots, u_{N_1+N_2+N_3n}) - h_i(\cdot,\cdot), u_{in}^{\approx} >_{\varrho}^{\tilde{}}]+$$

$$\sum_{i=N_1+N_2+1}^{N_1+N_2+N_3}[< D_t u_{in}, u_{in}^{\approx} >_{\varrho}^{\tilde{}} + \mu_i \mathcal{Q}_i^{\tilde{}}\,(u_{in}, u_{in}^{\approx}) - \mu_i(\lambda_1 - \varepsilon_i) < u_{in}, u_{in}^{\approx} >_{\varrho}^{\tilde{}}$$

$$- < f_i\,(\cdot,\cdot, u_{in}) + g_i(\cdot,\cdot, u_{1n}, \ldots, u_{N_1+N_2+N_3n}) - h_i(\cdot,\cdot), u_{in}^{\approx} >_{\varrho}^{\tilde{}}]$$

From (Q-1), (Q-2), (f-9), (f-10), (g-3), and (g-4) we see that

$$B : \mathbf{R}^{(N_1+N_2+N_3)(2n^2+n)} \to \mathbf{R}^{(N_1+N_2+N_3)(2n^2+n)}$$

is continuous in α where

$$B\,(\alpha) = (B_k^i\,(\alpha)_{k=1,\ldots,2n^2+n, i=1,\ldots,N_1+N_2+N_3}\,.$$

We propose to show that

$$\text{for } |\alpha| \text{ sufficiently large, } B(\alpha)\cdot\alpha > 0. \tag{4.46}$$

To accomplish this, we define

$$\varsigma_j(i) = \begin{cases} -1 & 1 \le j \le j_o(i)+j_1(i)-1 \\ 1 & j_o(i)+j_1(i) \le j \le n \end{cases}$$

for $i = 1, \ldots, N_1 + N_2$,

$$\varsigma_j(i) = 1$$

for $i = N_1 + N_2 + 1, \ldots, N_1 + N_2 + N_3$.

We then see from (4.40)-(4.43) that

$$u_{in}^{\approx} = \sum_{j=1}^{n} \varsigma_j(i)\, \hat{u}_{in}^c(j,0)\tilde{\varphi}_{j0}^c + \sum_{j=1}^{n} \varsigma_j(i) \sum_{k=1}^{n} \left[\hat{u}_{in}^c(j,k)\tilde{\varphi}_{jk}^c + \hat{u}_{in}^s(j,k)\tilde{\varphi}_{jk}^s\right], \tag{4.47}$$

$$u_{in} = \sum_{j=1}^{n} \hat{u}_{in}^c(j,0)\tilde{\varphi}_{j0}^c + \sum_{j=1}^{n}\sum_{k=1}^{n} \left[\hat{u}_{in}^c(j,k)\tilde{\varphi}_{jk}^c + \hat{u}_{in}^s(j,k)\tilde{\varphi}_{jk}^s\right],$$

and

$$D_t u_{in} = \sum_{j=1}^{n}\sum_{k=1}^{n} D_t \left[\hat{u}_{in}^c(j,k)\tilde{\varphi}_{jk}^c + \hat{u}_{in}^s(j,k)\tilde{\varphi}_{jk}^s\right]$$

$$= \sum_{j=1}^{n}\sum_{k=1}^{n} k \left[-\hat{u}_{in}^c(j,k)\tilde{\varphi}_{jk}^s + \hat{u}_{in}^s(j,k)\tilde{\varphi}_{jk}^c\right]$$

As a consequence, we have from (4.1), (4.25)(ii), and (4.47) that

$$< D_t u_{in}, u_{in}^{\approx} >_{\tilde{\varrho}} = 0 \quad \text{for i} = 1, \ldots, N_1 + N_2 + N_3. \tag{4.48}$$

Also, we see from (4.1) and (4.47), that

$$< u_{in}, u_{in}^{\approx} >_{\tilde{\varrho}} = \sum_{j=1}^{n} \varsigma_j(i) \left|\hat{u}_{in}^c(j,0)\right|^2 \tag{4.49}$$

$$+ \sum_{j=1}^{n} \varsigma_j(i) \sum_{k=1}^{n} \left|\hat{u}_{in}^c(j,k)\right|^2 + \left|\hat{u}_{in}^s(j,k)\right|^2$$

and from (4.3), (4.7), and the bilinearity of $\mathcal{L}^{\tilde{}}(\cdot,\cdot)$ that

$$\mathcal{L}^{\tilde{}}(u_{in}, u_{in}^{\approx}) = \sum_{j=1}^{n} \varsigma_j(i)\, \lambda_j \left|\hat{u}_{in}^c(j,0)\right|^2 \tag{4.50}$$

$$+ \sum_{j=1}^{n} \varsigma_j(i) \sum_{k=1}^{n} \left[\left|\hat{u}_{in}^c(j,k)\right|^2 + \left|\hat{u}_{in}^s(j,k)\right|^2\right] \lambda_j.$$

Next from (4.38) and (4.39), we see that for $n > n_0^*$,

$$\lambda_j - (\lambda_{j_o(i)} + \frac{1}{n}) \geq \frac{1}{n} \quad \text{for } j \geq j_o(i) + j_1(i) \text{ and } i = 1, \ldots, N_1.$$

This fact in conjunction with (4.49) and (4.50) then gives us that

$$\mathcal{L}^{\tilde{}}(u_{in}, u_{in}^{\approx}) - (\lambda_{j_o(i)} + \frac{1}{n}) < u_{in}, u_{in}^{\approx} >_{\varrho}^{\tilde{}} \geq \frac{1}{n} < u_{in}, u_{in} >_{\varrho}^{\tilde{}}$$

for $i = 1, \ldots, N_1$. Likewise, we also see from (4.38) and (4.39) that for $n > n_0^*$,

$$\lambda_j - (\lambda_{j_o(i)} + \varepsilon_i) \geq \frac{1}{n} \quad \text{for } j \geq j_o(i) + j_1(i) \text{ and } i = N_1 + 1, \ldots, N_1 + N_2,$$

and obtain, once again, using (4.49) and (4.50) that

$$\mathcal{L}^{\tilde{}}(u_{in}, u_{in}^{\approx}) - (\lambda_{j_o(i)} + \varepsilon_i) < u_{in}, u_{in}^{\approx} >_{\varrho}^{\tilde{}} \geq \frac{1}{n} < u_{in}, u_{in} >_{\varrho}^{\tilde{}}$$

for $i = N_1 + 1, \ldots, N_1 + N_2$, and $n > n_0^*$. Observing from (4.42) and (4.43) that $u_{in} = u_{in}^{\approx}$ for $i = N_1 + N_2 + 1, \ldots, N_1 + N_2 + N_3$ and also that $\varsigma_j(i) = 1$ for j=1,...,n, we see from (4.49), (4.50), (4.38), and (4.39) that

$$\mathcal{L}^{\tilde{}}(u_{in}, u_{in}^{\approx}) - (\lambda_1 - \varepsilon_i) < u_{in}, u_{in}^{\approx} >_{\varrho}^{\tilde{}} \geq \frac{1}{n} < u_{in}, u_{in} >_{\varrho}^{\tilde{}}$$

for $i = N_1 + N_2 + 1, \ldots, N_1 + N_2 + N_3$ and $n > n_0^*$. We conclude from the above that

$$\begin{aligned} &\sum_{i=1}^{N_1} \mu_i [\mathcal{L}^{\tilde{}}(u_{in}, u_{in}^{\approx}) - (\lambda_{j_o(i)} + \frac{1}{n}) < u_{in}, u_{in}^{\approx} >_{\varrho}^{\tilde{}}] + \\ &\sum_{i=N_1+1}^{N_1+N_2} \mu_i [\mathcal{L}^{\tilde{}}(u_{in}, u_{in}^{\approx}) - (\lambda_{j_o(i)} + \varepsilon_i) < u_{in}, u_{in}^{\approx} >_{\varrho}^{\tilde{}}] + \\ &\sum_{i=N_1+N_2+1}^{N_1+N_2+N_3} \mu_i [\mathcal{L}^{\tilde{}}(u_{in}, u_{in}^{\approx}) - (\lambda_1 - \varepsilon_i) < u_{in}, u_{in}^{\approx} >_{\varrho}^{\tilde{}}] \\ &\geq \frac{1}{n} \sum_{i=1}^{N_1+N_2+N_3} \mu_i < u_{in}, u_{in} >_{\varrho}^{\tilde{}} \end{aligned} \tag{4.51}$$

We set

$$\mu_0 = min\,(\mu_1, \ldots, \mu_{N_1+N_2+N_3}) > 0$$

and observe from(4.45), (4.48), and (4.51) that

$$B(\alpha)\cdot\alpha \geq \frac{\mu_0}{n}\sum_{i=1}^{N_1+N_2+N_3} < u_{in}, u_{in} >_{\varrho}^{\tilde{}} \tag{4.52}$$

$$+\sum_{i=1}^{N_1+N_2+N_3} \mu_i \left[\mathcal{Q}_i^{\tilde{}}\,(u_{in}, u_{in}^{\approx}) - \mathcal{L}^{\tilde{}}\,(u_{in}, u_{in}^{\approx})\right]$$

$$-\sum_{i=1}^{N_1+N_2+N_3} < f_i\,(\cdot,\cdot,u_{in}) + g_i(\cdot,\cdot,u_{1n},\ldots,u_{N_1+N_2+N_3n}) - h_i(\cdot,\cdot), u_{in}^{\approx} >_{\varrho}^{\tilde{}}\,.$$

Next, we see from (4.40) and from (4.43) that

$$< u_{in}^{\approx}, u_{in}^{\approx} >_{\varrho}^{\tilde{}} = < u_{in}, u_{in} >_{\varrho}^{\tilde{}} = \sum_{k=1}^{2n^2+n} \left|\alpha_k^i\right|^2 \tag{4.53}$$

Also, we see (4.3), (4.7), (4.47),and the bilinearity $\mathcal{L}^{\tilde{}}\,(\cdot,\cdot)$ that

$$\mathcal{L}^{\tilde{}}\,(u_{in}^{\approx}, u_{in}^{\approx}) = \sum_{j=1}^{n} \lambda_j \left|\hat{u}_{in}^{c}(j,0)\right|^2 + \sum_{j=1}^{n}\sum_{k=1}^{n}\left[\left|\hat{u}_{in}^{c}(j,k)\right|^2 + \left|\hat{u}_{in}^{s}(j,k)\right|^2\right]\lambda_j$$

Consequently,

$$\left|\mathcal{L}^{\tilde{}}\,(u_{in}^{\approx}, u_{in}^{\approx})\right|^{\frac{1}{2}} \leq \sqrt{\lambda_n}\{\sum_{k=1}^{2n^2+n}\left|\alpha_k^i\right|^2\}^{\frac{1}{2}} \tag{4.54}$$

Now from the definition of α (see two lines above (4.44)),

$$|\alpha|^2 = \sum_{i=1}^{N_1+N_2+N_3}\sum_{k=1}^{2n^2+n}\left|\alpha_k^i\right|^2.$$

Also, from the fact that is Q_i is #-#$\tilde{\mathrm{H}}$ related to L, we see from (2.3)(ii) that there are positive constants K_1 and K_2 such that.

$$\left|\mathcal{Q}_i^{\tilde{}}(u_{in}, u_{in}^{\approx}) - \mathcal{L}^{\tilde{}}\,(u_{in}, u_{in}^{\approx})\right| \leq K_1\left|\mathcal{L}^{\tilde{}}\,(u_{in}^{\approx}, u_{in}^{\approx})\right|^{\frac{1}{2}} + K_2$$

for $i = 1,\ldots,N_1+N_2+N_3$. So we conclude from (4.52)-(4.54) that

$$B(\alpha)\cdot\alpha \geq \frac{\mu_0}{n}|\alpha|^2 - [K_1\sqrt{\lambda_n}\,|\alpha| + K_2]\mu_n^*\,(N_1+N_2+N_3) \tag{4.55}$$

$$-\sum_{i=1}^{N_1+N_2+N_3} < f_i(\cdot,\cdot,u_{in}) + g_i(\cdot,\cdot,u_{1n},\ldots,u_{N_1+N_2+N_3n}) - h_i(\cdot,\cdot), u_{in}^{\approx} >_{\varrho}^{\tilde{}}$$

where $\mu_n^* = max\,(\mu_1,\ldots,\mu_{N_1+N_2+N_3})$.

Next, we see from (f-10) and (g-4) that there is a positive K_3 such that

$$\left|< f_i(\cdot,\cdot,u_{in}) + g_i(\cdot,\cdot,u_{1n},\ldots,u_{N_1+N_2+N_3n}) - h_i(\cdot,\cdot), u_{in}^{\approx} >_{\varrho}^{\tilde{}}\right|$$

$$\leq K_3 \left|< u_{in}^{\approx}, u_{in}^{\approx} >_{\varrho}^{\tilde{}}\right|^{\frac{1}{2}}$$

for $i = 1,...,N_1 + N_2 + N_3$, and hence from (4.53) and (4.55) that

$$B(\alpha)\cdot\alpha \geq \frac{\mu_0}{n}|\alpha|^2 - [K_1\sqrt{\lambda_n}\,|\alpha| + K_2]\mu_n^*\,(N_1+N_2+N_3)$$

$$-K_3\,|\alpha|\,(N_1+N_2+N_3)\,.$$

Since n is fixed and $\mu_0 > 0$, we see from this last inequality that there exists $s_o^* > 0$ such that

$$B(\alpha)\cdot\alpha \geq \frac{\mu_0}{2n}|\alpha|^2 \quad \text{for } |\alpha| \geq s_o^*. \tag{4.56}$$

Consequently, (4.46) is established.

As we already have observed,

$$B\,(\alpha) = (B_k^i\,(\alpha))_{k=1,\ldots,2n^2+n,i=1,\ldots,N_1+N_2+N_3}$$

is a continuous map of $\mathbf{R}^{(N_1+N_2+N_3)(2n^2+n)}$ into itself. Therefore, it follows from [Ke, p. 219] or [Ni, p. 18] that there exists

$$\alpha^* = \left(\alpha_k^{i*}\right)_{k=1,\ldots,2n^2+n,i=1,\ldots,N_1+N_2+N_3} \in \mathbf{R}^{(2n^2+n)(N_1+N_2+N_3)}$$

such that

$$B_k^i\,(\alpha^*) = 0 \text{ for i} = 1,\ldots,N_1+N_2+N_3, \text{k}=1,\ldots,2n^2+n. \tag{4.57}$$

Using (4.43) and setting

$$u_{in}^* = \sum_{k=1}^{2n^2+n} \alpha_k^{i*}\psi_k$$

we therefore see from (4.44) and (4.57) that

$$< D_t u_{in}^*, \psi_k >_{\varrho}^{\tilde{}} + \mu_i \tilde{Q}_i\,(u_{in},\psi_k)$$

$$= \mu_i(\lambda_{j_o(i)} + \frac{1}{n}) < u_{in},\psi_k >_{\varrho}^{\tilde{}} + < f_i(\cdot,\cdot,u_{in}) + \tag{4.58}$$

$$g_i(\cdot,\cdot,u_{1n},\ldots,u_{N_1+N_2+N_3n}) - h_i(\cdot,\cdot),\psi_k >_{\varrho}^{\tilde{}} \qquad i=1,\ldots,N_1,$$

$$= \mu_i(\lambda_{j_o(i)}+\varepsilon_i) < u_{in},\psi_k >_{\varrho}^{\tilde{}} + < f_i(\cdot,\cdot,u_{in}) + g_i(\cdot,\cdot,u_{1n},$$

$$\ldots,u_{N_1+N_2+N_3n}) - h_i(\cdot,\cdot),\psi_k >_{\varrho}^{\tilde{}} \; i=N_1+1,\ldots,N_1+N_2,$$

$$= \mu_i(\lambda_1-\varepsilon_i) < u_{in},\psi_k >_{\varrho}^{\tilde{}} + < f_i(\cdot,\cdot,u_{in}) + g_i(\cdot,\cdot,u_{1n},\ldots,$$

$$u_{N_1+N_2+N_3n}) - h_i(\cdot,\cdot),\psi_k >_{\varrho}^{\tilde{}} \quad i=N_1+N_2+1,\ldots,N_1+N_2+N_3$$

for k=1,..., $2n^2+n$.

Since every $v \in S_n^{\ddagger}$ is of the form $v=\sum_{k=1}^{2n^2+n}\beta_k\psi_k$, we see from (4.58) and the linearity of $\mathcal{Q}_i^{\tilde{}}(\cdot,\cdot)$ in the second variable that (4.37) does indeed hold for $(u_{1n}^*,\ldots,u_{N_1+N_2+N_3n}^*)$ and every $v\in S_n^{\ddagger}$. Since $u_{in}^*\in S_n^{\ddagger}$ for $i=1,\ldots,N_1+N_2+N_3$, the proof of the lemma is complete.

2.5 Proof of Theorem 1

Without loss of generality, we assume from the start that $j_o>1$. (A similar but easier proof prevails in case $j_o=1$.) We also assume that λ_{j_o} is an eigenvalue of L of multiplicity j_1 and that $\lambda_{j_o+j_1}$ is an eigenvalue of L of multiplicity j_2. We start out the proof by invoking Lemma 4 with $n_1=j_o+j_1+j_2$. Hence we have $\{u_n\}_{n=n_1}^{\infty}\subset\tilde{H}$ with $u_n\in S_n^{\ddagger}$ such that

$$< D_t u_n, v >_{\varrho}^{\tilde{}} + \mathcal{Q}^{\tilde{}}(u_n,v) = (\lambda_{j_o}+\gamma n^{-1}) < u_n, v >_{\varrho}^{\tilde{}} \tag{5.1}$$

$$+(1-n^{-1}) < f(\cdot,u_n) + g(\cdot,\cdot,u_n), v >_{\varrho}^{\tilde{}} \quad \forall v\in S_n^{\ddagger}.$$

We claim there is a constant K such that

$$\|u_n\|_{\tilde{H}} \le K \quad \forall n \ge n_1. \tag{5.2}$$

Suppose that (5.2) does not hold. Then without loss in generality and for ease of notation, we can assume that

$$\lim_{n\to\infty}\|u_n\|_{\tilde{H}} = \infty. \tag{5.3}$$

We shall show that (5.3) leads to a contradiction.

In order to accomplish this, we first show that (5.1) and (5.3) together imply that

$$\lim_{n\to\infty}\|u_n\|_{\tilde{L}^2_{\varrho}} = \infty \tag{5.4}$$

where $\tilde{L}^2_\varrho = L^2_\varrho\left(\tilde{\Omega}\right)$.

Suppose that (5.4) is false. Then without loss in generality, we can assume there is a constant $K_1 > 0$ and that

$$\|u_n\|_{\tilde{L}^2_\varrho} \leq K_1 \quad \text{for } n \geq n_1. \tag{5.5}$$

We take $v = u_n$ in (5.1) and observe from (4.23) and (4.25)(i) that $< D_t u_n, u_n >_{\tilde{\varrho}} = 0$. Also, we observe from (f-2) and (g-2) and Schwarz's inequality that

$$\left|< f(\cdot, u_n) + g(\cdot, \cdot, u_n), u_n >_{\tilde{\varrho}}\right| \leq K_2 \|u_n\|^2_{\tilde{L}^2_\varrho} + K_2 \|u_n\|_{\tilde{L}^2_\varrho}$$

for $n \geq n_1$ where K_2 is a constant independent of n. Therefore it follows from (5.1) with $v = u_n$ that there is a constant K_3 such that

$$\tilde{\mathcal{Q}}(u_n, u_n) \leq K_3 \|u_n\|^2_{\tilde{L}^2_\varrho} + K_3 \|u_n\|_{\tilde{L}^2_\varrho} \quad \forall n \geq n_1. \tag{5.6}$$

Next, we observe from (4.25)(i) that $D_t u_n \in S^\ddagger_n$. Also, we see from Remark 2 (which is between Lemma 3 and Lemma 4) that

$$\tilde{\mathcal{Q}}(u_n, D_t u_n) = 0 \quad \text{for } n \geq n_1. \tag{5.7}$$

Furthermore, we see that

$$\int_T u_n(x,t) D_t u_n(x,t) dt = 2^{-1} \int_T D_t u_n^2(x,t) dt = 0$$

for $x \in \Omega$ where $T = (-\pi, \pi)$. Consequently, (as we have observed before)

$$< u_n, D_t u_n >_{\tilde{\varrho}} = 0 \quad \forall n \geq n_1.$$

Hence we obtain from (5.1) with $v = D_t u_n$ that

$$\|D_t u_n\|^2_{\tilde{L}^2_\varrho} \leq | < f(\cdot, u) + g(\cdot, \cdot, u_n), D_t u_n >_{\tilde{\varrho}} | \tag{5.8}$$

$$\leq K_4 \|u_n\|_{\tilde{L}^2_\varrho} \|D_t u_n\|_{\tilde{L}^2_\varrho} + K_4 \|D_t u_n\|_{\tilde{L}^2_\varrho}$$

where we use (f-2) and (g-2) once again and K_4 is independent of n.

From (Q-3), (I,1.5)(iii)), and (1.10)(ii), we also see that there is a positive constant C_4 such that

$$C_4\left[\sum_{i=1}^N \int_{\tilde{\Omega}} p_i |D_i u_n|^2 + \int_{\tilde{\Omega}} q u_n^2\right] \leq \tilde{\mathcal{Q}}(u_n, u_n) \tag{5.9}$$

Therefore it follows from (5.6), (5.8), (5.9), (1.2) and (5.5) that

$$\|u_n\|_{\tilde{H}} \leq K_5 \quad \forall\, n \geq n_1.$$

This fact contradicts (5.3) and this contradiction was obtained on the basis of (5.5). Hence (5.5) is false and (5.4) is indeed true.

Next, we see from (5.4), (5.6), (5.8), and (5.9) that

$$\exists\, K_6 > 0 \ s.t \ \|u_n\|_{\tilde{H}} \leq K_6 \ \|u_n\|_{\tilde{L}^2_\varrho} \quad \forall\, n \geq n_1. \tag{5.10}$$

Also, we write

$$u_n = u_{n1} + u_{n2} + u_{n3} + u_{n4} \quad where \tag{5.11}$$

$$u_{n1} = \sum_{j=1}^{j_o-1} \hat{u}^c_n(j,0)\tilde{\varphi}^c_{j0} + \sum_{j=1}^{j_o-1}\sum_{k=1}^{n}[\hat{u}^c(j,k)\tilde{\varphi}^c_{jk} + \hat{u}^s(j,k)\tilde{\varphi}^s_{jk}],$$

$$u_{n2} = \sum_{j=j_o}^{j_o+j_1-1} \hat{u}^c_n(j,0)\tilde{\varphi}^c_{j0} + \sum_{j=j_o}^{j_o+j_1-1}\sum_{k=1}^{n}[\hat{u}^c(j,k)\tilde{\varphi}^c_{jk} + \hat{u}^s(j,k)\tilde{\varphi}^s_{jk}],$$

$$u_{n3} = \sum_{j=j_o+j_1}^{j_o+j_1+j_2-1} \hat{u}^c_n(j,0)\tilde{\varphi}^c_{j0} + \sum_{j=j_o+j_1}^{j_o+j_1+j_2-1}\sum_{k=1}^{n}[\hat{u}^c(j,k)\tilde{\varphi}^c_{jk} + \hat{u}^s(j,k)\tilde{\varphi}^s_{jk}],$$

$$u_{n4} = \sum_{j=j_o+j_1+j_2}^{n} \hat{u}^c_n(j,0)\tilde{\varphi}^c_{j0} + \sum_{j=j_o+j_1+j_2}^{n}\sum_{k=1}^{n}[\hat{u}^c(j,k)\tilde{\varphi}^c_{jk} + \hat{u}^s(j,k)\tilde{\varphi}^s_{jk}].$$

We claim the following:

$$(i)\ \lim_{n\to\infty}[\|u_{n1}\|^2_{\tilde{L}^2_\varrho} + \|u_{n4}\|^2_{\tilde{L}^2_\varrho}] / \|u_n\|^2_{\tilde{L}^2_\varrho} = 0; \tag{5.12}$$

$$(ii)\ \lim_{n\to\infty} \|u_{n4}\|_{\tilde{H}} \,/\, \|u_n\|_{\tilde{L}^2_\varrho} = 0.$$

To establish this claim, we define $u_n^{\approx}$ as before (See (4.31).); so

$$u_n^{\approx} = -u_{n1} - u_{n2} + u_{n3} + u_{n4}. \tag{5.13}$$

Since $< D_t u_{ni}, u_{nk} >_{\tilde{\varrho}} = 0$ for $i \neq k$, we see from (4.25)(ii) that

$$< D_t u_n, u_n^{\approx} >_{\tilde{\varrho}} = 0. \tag{5.14}$$

Consequently, we obtain from (5.1) with $v = u_n^{\approx}$ that

$$\mathcal{L}^{\tilde{}}(u_n, u_n^{\approx}) - (\lambda_{j_o} + \gamma) < u_n, u_n^{\approx} >_{\tilde{\varrho}} \tag{5.15}$$

$$= (1 - n^{-1}) < f(\cdot, u_n) - \gamma u_n, u_n^{\approx} >_{\varrho}^{\tilde{}} + < g(\cdot, \cdot, u_n), u_n^{\approx} >_{\varrho}^{\tilde{}}$$

$$\mathcal{L}^{\tilde{}}(u_n, u_n^{\approx}) - \mathcal{Q}^{\tilde{}}(u_n, u_n^{\approx}).$$

Now from (4.31) and the line below, we see that the left-hand side of (5.15) is

$$\sum_{j=1}^{j_o+j_1-1} (\lambda_{j_o} + \gamma - \lambda_j) |\hat{u}_n^c(j,0)|^2 + \sum_{j=j_o+j_1}^{n} (\lambda_j - \lambda_{j_o} - \gamma) |\hat{u}_n^c(j,0)|^2$$

$$+ \sum_{j=1}^{j_o+j_1-1} (\lambda_{j_o} + \gamma - \lambda_j) \sum_{k=1}^{n} [|\hat{u}_n^c(j,k)|^2 + |\hat{u}_n^s(j,k)|^2]$$

$$+ \sum_{j=j_o+j_1}^{n} (\lambda_j - \lambda_{j_o} - \gamma) \sum_{k=1}^{n} [|\hat{u}_n^c(j,k)|^2 + |\hat{u}_n^s(j,k)|^2].$$

But from (1,2.10) we also see that $\lambda_{j_o} + \gamma = \lambda_{j_o+j_1} - \gamma$. Therefore, it follows from this last computation that the left-hand side of (5.15) majorizes

$$\gamma \|u_n\|^2_{\tilde{L}^2_\varrho} + \sum_{j=1}^{j_o-1} (\lambda_{j_o} - \lambda_j) |\hat{u}_n^c(j,0)|^2 + \sum_{j=j_o+j_1+j_2}^{n} (\lambda_j - \lambda_{j_o+j_1}) |\hat{u}_n^c(j,0)|^2$$

$$+ \sum_{j=1}^{j_o-1} (\lambda_{j_o} - \lambda_j) \sum_{k=1}^{n} [|\hat{u}_n^c(j,k)|^2 + |\hat{u}_n^s(j,k)|^2] \tag{5.16}$$

$$+ \sum_{j=j_o+j_1+j_2}^{n} (\lambda_j - \lambda_{j_o+j_1}) \sum_{k=1}^{n} [|\hat{u}_n^c(j,k)|^2 + |\hat{u}_n^s(j,k)|^2].$$

On the other hand, given $\varepsilon > 0$, we see from (f-2), (g-2), (4.4), (5.10), and (1.10) that the right-hand side of (5.15) is majorized by

$$[(\gamma + \varepsilon) \|u_n\|_{\tilde{L}^2_\varrho} + \|f_0\|_{\tilde{L}^2_\varrho} + \|g_\varepsilon\|_{\tilde{L}^2_\varrho}] \|u_n^{\approx}\|_{\tilde{L}^2_\varrho} + \eta_n \|u_n\|_{\tilde{L}^2_\varrho} \|u_n^{\approx}\|_{\tilde{L}^2_\varrho} \tag{5.17}$$

where $\{\eta_n\}_{n=n_1}^{\infty}$ is a decreasing sequence of positive real numbers with $\eta_n \to 0$. Hence, (5.17) majorizes (5.16).

Next, we see from (5.11) that (5.16) in turn majorizes

$$\gamma \|u_n\|^2_{\tilde{L}^2_\varrho} + (\lambda_{j_o} - \lambda_{j_o-1}) \|u_{n1}\|^2_{\tilde{L}^2_\varrho} + (\lambda_{j_o+j_1+j_2} - \lambda_{j_o+j_1}) \|u_{n4}\|^2_{\tilde{L}^2_\varrho}. \tag{5.18}$$

Therefore, (5.17) majorizes (5.18). But $\|u_n\|^2_{\tilde{L}^2_\varrho} = \|u_n^{\approx}\|^2_{\tilde{L}^2_\varrho}$, and we see on cancelling $\gamma\, \|u_n\|^2_{\tilde{L}^2_\varrho}$, and then dividing both (5.17) and (5.18) by $\|u_n\|^2_{\tilde{L}^2_\varrho}$ and passing to the limit, that

$$\lim_{n\to\infty} \left[(\lambda_{j_o} - \lambda_{j_o-1}) \|u_{n1}\|^2_{\tilde{L}^2_\varrho} + (\lambda_{j_o+j_1+j_2} - \lambda_{j_o+j_1}) \|u_{n4}\|^2_{\tilde{L}^2_\varrho} \right] / \|u_n\|^2_{\tilde{L}^2_\varrho} \le \varepsilon$$

However, $\varepsilon > 0$ is arbitrary; so this last limit is actually equal to zero. Also both $(\lambda_{j_o} - \lambda_{j_o-1})$ and $(\lambda_{j_o+j_1+j_2} - \lambda_{j_o+j_1})$ are strictly positive numbers. So we see that claim (i) in (5.12) is indeed true.

To establish claim (ii) in (5.12), we first observe that

$$\exists\, \delta > 0 \text{ s.t. } \lambda_j - \lambda_{j_o+j_1} \ge \delta\, \lambda_j \quad \forall\, j \ge j_o + j_1 + j_2. \tag{5.19}$$

Next, we return to (5.16) and see from (5.19) that (5.16) majorizes

$$\gamma\, \|u_n\|^2_{\tilde{L}^2_\varrho} + \sum_{j=j_o+j_1+j_2}^{n} \delta\lambda_j\, |\hat{u}_n^c(j,0)|^2 +$$
$$\sum_{j=j_o+j_1+j_2}^{n} \delta\lambda_j \sum_{k=1}^{n} [|\hat{u}_n^c(j,k)|^2 + |\hat{u}_n^s(j,k)|^2],$$

which from (5.11) is in turn equal to

$$\gamma\, \|u_n\|^2_{\tilde{L}^2_\varrho} + \delta\mathcal{L}^{\tilde{}}\, (u_{n4}, u_{n4})\,. \tag{5.20}$$

What this means, in particular, since (5.17) majorizes (5.16) is that (5.17) majorizes (5.20). Hence, cancelling $\gamma\, \|u_n\|^2_{\tilde{L}^2_\varrho}$ from both (5.17) and (5.20) and then dividing both by $\|u_n\|^2_{\tilde{L}^2_\varrho}$ and passing to the limit, we obtain that

$$\lim_{n\to\infty} \delta\mathcal{L}^{\tilde{}}\, (u_{n4}, u_{n4}) \,/\, \|u_n\|^2_{\tilde{L}^2_\varrho} \le \varepsilon.$$

But $\varepsilon > 0$ is arbitrary. Therefore

$$\lim_{n\to\infty} \mathcal{L}^{\tilde{}}\, (u_{n4}, u_{n4}) \,/\, \|u_n\|^2_{\tilde{L}^2_\varrho} = 0. \tag{5.21}$$

Consequently, claim (ii) in (5.12) will follow from claim (i), (4.4), and (5.21) plus the fact that $\|D_t u_{n4}\|^2_{\tilde{L}^2_\varrho} \le \|D_t u_n\|^2_{\tilde{L}^2_\varrho}$, once we show that the following holds:

$$\lim_{n\to\infty} \|D_t u_n\|^2_{\tilde{L}^2_\varrho} \,/\, \|u_n\|^2_{\tilde{L}^2_\varrho} = 0. \tag{5.22}$$

To establish (5.22), we first observe from (1.8) that if v$\in \tilde{C}^{1b}_{p,\varrho}$, then both v and $D_t v$ are in $\tilde{C}^{1}_{p,\varrho}$. So for v$\in \tilde{C}^{1b}_{p,\varrho}$, on setting $F(x,t) = \int_0^{v(x,t)} f(x,s)\, ds$, we see that $F(x,t)$ is periodic of period 2π in t for a.e. x$\in \Omega$. Also,

$$D_t F(x,t) = f(x, v(x,t))\; D_t v(x,t) \quad for\ a.e.\ x \in \Omega.$$

As a consequence,

$$\langle f(\cdot, v), D_t v\rangle^{\tilde{}}_{\varrho} = \int_\Omega [\int_0^{2\pi} f(x, v(x,t))\; D_t v(x,t)\, dt] dx = 0 \qquad (5.23)$$

for v$\in \tilde{C}^{1b}_{p,\varrho}$. Now it is easy to see that $u_n \in S^{\ddagger}_{n}$ implies that there

$$\exists \{v_k\}_{k=1}^{\infty} \text{ with } v_k \in \tilde{C}^{1b}_{p,\varrho} \text{ s.t. } \|v_k - u_n\|_{\tilde{H}} \to 0.$$

Therefore, we conclude from (5.23) and $(f-2)$that

$$\langle f(\cdot, u_n), D_t u_n\rangle^{\tilde{}}_{\varrho} = 0 \quad for\ n \geq n_1.$$

In a similar vein we obtain from $(Q-6)$ (see Remark 2) that

$$\mathcal{Q}^{\tilde{}}(u_n, D_t u_n) = 0 \quad for\ n \geq n_{1.}$$

Since we already know that $\langle u_n, D_t u_n\rangle^{\tilde{}}_{\varrho} = 0$, it follows from these last two facts on replacing v by $D_t u_n$ in (5.1) that

$$\langle D_t u_n, D_t u_n\rangle^{\tilde{}}_{\varrho} = (1 - n^{-1}) < g(\cdot, \cdot, u_n), D_t u_n >^{\tilde{}}_{\varrho} \quad for\ n \geq n_1$$

But then Schwarz's inequality implies that

$$\|D_t u_n\|_{\tilde{L}^2_\varrho} \leq \|g(\cdot, \cdot, u_n)\|_{\tilde{L}^2_\varrho},$$

and (5.21) follows immediately from (5.4) and $(g-2)$ applied to this last inequality. Therefore, claim (*ii*) in (5.12) is completely established.

Continuing with our development to show that (5.3) leads to a contradiction, we set

$$W_n = \frac{u_n}{\|u_n\|_{\tilde{L}^2_\varrho}} \text{ and } W_{ni} = \frac{u_{ni}}{\|u_n\|_{\tilde{L}^2_\varrho}} \quad \text{for i=1,\ldots,4} \qquad (5.24)$$

and observe from (4.6) that $\mathcal{L}^{\tilde{}}_1(W_{ni}, W_{ni}) \leq \mathcal{L}^{\tilde{}}_1(W_n, W_n)$ for i=1,. . . ,4. Also, it follows from (4.4) that

$$\exists\, K_7\ s.t.\ \mathcal{L}^{\tilde{}}_1(W_n, W_n) \leq K_7 \|W_n\|^2_{\tilde{H}} \quad for\ n \geq n_1.$$

Consequently, we obtain from (4.4), (5.10), and (5.22) that

$$\exists\, \mathrm{K}_8 \; s.t. \; \|W_n\|_{\tilde{H}} \leq \mathrm{K}_8 \text{ and } \|W_{ni}\|_{\tilde{H}} \leq \mathrm{K}_8 \tag{5.25}$$

for i=1,...,4 and $n \geq n_1$.In the above, we also used the fact that

$$\|D_t W_{ni}\|^2_{\tilde{L}^2_\varrho} \leq \|D_t W_n\|^2_{\tilde{L}^2_\varrho} \qquad \text{for } i = 1, \ldots, 4.$$

As a consequence of the set of inequalities in (5.25), we obtain, in particular, from the fact that Q is #H̃-related to L that

$$\lim_{n\to\infty} \left[\frac{\mathcal{Q}^{\tilde{}}(u_n, W_{ni}) - \mathcal{L}^{\tilde{}}(u_n, W_{ni})}{\|u_n\|_{\tilde{L}^2_\varrho}} \right] = 0 \quad \text{for i=1,\ldots,4.} \tag{5.26}$$

Another consequence of this same set of inequalities when joined with Lemma 3 is the following (where we have used a full sequence instead of a subsequence for ease of notation):

$$\begin{aligned} \exists\, W \;&\in\; \tilde{H} \text{ such that} \\ (i) \lim_{n\to\infty} \|W_n - W\|_{\tilde{L}^2_\varrho} \;&=\; 0; \end{aligned} \tag{5.27}$$

$$(ii) \lim_{n\to\infty} W_n(x,t) = W(x) \text{ a.e. in } \tilde{\Omega};$$

$$(iii) \lim_{n\to\infty} < W_n, v >_{\tilde{H}} = < W, v >_{\tilde{H}} \; \forall v \in \tilde{H}.$$

Next, we observe that

$$\hat{W}^c(j,k) = 0 \text{ and } \hat{W}^s(j,k) = 0 \text{ for } j \geq j_o + j_1 + j_2 \text{ and } \forall k. \tag{5.28}$$

To see this fact, we observe from (5.12) that $lim_{n\to\infty} \|W_{n4}\|_{\tilde{L}^2_\varrho} = 0$. Hence, it follows that

$$< W_n, \tilde{\varphi}^c_{jk} >_{\varrho}^{\tilde{}} = < W_{n4}, \tilde{\varphi}^c_{jk} >_{\varrho}^{\tilde{}} \to 0 \text{ for } j \geq j_o + j_1 + j_2,$$

and the first part of (5.28) follows from (5.27) (i). In a similar manner, we establish the second part. Also, we observe from (5.12) that $lim_{n\to\infty} \|W_{n1}\|_{\tilde{L}^2_\varrho} = 0$. So we obtain in a similar vein

$$\hat{W}^c(j,k) = 0 \text{ and } \hat{W}^s(j,k) = 0 \text{ for } 1 \leq j \leq j_o - 1 \text{ and } \forall k. \tag{5.29}$$

Next, suppose k$\geq$ 1 and $j_o \leq j \leq j_o + j_1 + j_2 - 1$, and also recall from (5.22) that $\lim_{n\to\infty} \|D_t W_n\|^2_{\tilde{L}^2_\varrho} = 0$. Then from (4.2) we have

$$\begin{aligned} k\hat{W}^c(j,k) &= \lim_{n\to\infty} \int_{\tilde{\Omega}} W_n(x,t)\, D_t\tilde{\varphi}^s_{jk}(x,t)\, \varrho(x)\, dxdt \\ &= -\lim_{n\to\infty} \int_{\tilde{\Omega}} D_t W_n(x,t)\ \tilde{\varphi}^s_{jk}(x,t)\, \varrho(x)\, dxdt \\ &= 0. \end{aligned}$$

A similar situation prevails for $k\hat{W}^s(j,k)$. So we have

$$\hat{W}^c(j,k) = 0 \text{ and } \hat{W}^s(j,k) = 0 \text{ for } k \geq 1 \text{ and } j_o \leq j \leq j_o + j_1 + j_2 - 1.$$

As a consequence of this last fact joined with (5.27) and (5.28), we have that

$$\begin{aligned} W(x) &= W_2(x) + W_3(x) \text{ where} \qquad (5.30)\\ W_2(x) &= \sum_{j=j_o}^{j_o+j_1-1} \hat{W}^c(j,0)\tilde{\varphi}^c_{j0} \text{ and } W_3(x) = \sum_{j=j_o+j_1}^{j_o+j_1+j_2-1} \hat{W}^c(j,0)\tilde{\varphi}^c_{j0}. \end{aligned}$$

Next, we observe from (5.24) and orthogonality that

$$\begin{aligned} \|W_n - W\|^2_{\tilde{L}^2_\varrho} &= \|W_{n1} + W_{n2} + W_{n3} + W_{n4} - W_2 - W_3\|^2_{\tilde{L}^2_\varrho} \\ &= \|W_{n1}\|^2_{\tilde{L}^2_\varrho} + \|W_{n2} - W_2\|^2_{\tilde{L}^2_\varrho} \\ &\quad + \|W_{n3} - W_3\|^2_{\tilde{L}^2_\varrho} + \|W_{n4}\|^2_{\tilde{L}^2_\varrho}. \end{aligned}$$

Since by (5.27)(i), the left-hand side of the above equality goes to zero as n$\to \infty$, so does the right-hand side. Since all the terms on the right-hand side are nonnegative, we conclude that both

$$\lim_{n\to\infty} \|W_{n2} - W_2\|^2_{\tilde{L}^2_\varrho} = 0 \text{ and } \lim_{n\to\infty} \|W_{n3} - W_3\|^2_{\tilde{L}^2_\varrho} = 0. \qquad (5.31)$$

Continuing with our development to show that (5.3) leads to a contradiction, we put W_{ni} in place of v in (5.1) and obtain

$$\begin{aligned} \tilde{\mathcal{L}}(u_n, W_{ni}) &= (\lambda_{j_o} + \gamma n^{-1}) < u_n, W_{ni} >_{\tilde{\varrho}} \qquad (5.32)\\ &+ (1 - n^{-1}) < f(\cdot, u_n) + g(\cdot,\cdot,u_n), W_{ni} >_{\tilde{\varrho}} \\ &+ \tilde{\mathcal{L}}(u_n, W_{ni}) - \tilde{\mathcal{Q}}(u_n, W_{ni}), \end{aligned}$$

where we have used the fact from (5.24) that

$$< D_t u_n, W_{ni} >_{\varrho}^{\tilde{}} = < D_t u_{ni}, u_{ni} >_{\varrho}^{\tilde{}} / \|u_{ni}\|_{\tilde{L}^2_{\varrho}} = 0.$$

Now from (5.11) and (5.24), we see that

$$\mathcal{L}^{\tilde{}}(u_n, W_{n3}) = \lambda_{j_o+j_1} < u_n, W_{n3} >_{\varrho}^{\tilde{}}.$$

Also from (5.24), Schwarz's inequality,and (g-2), we obtain that

$$\lim_{n\to\infty} | < g(\cdot,\cdot,u_n), W_{ni} >_{\varrho}^{\tilde{}} | / \|u_n\|_{\tilde{L}^2_{\varrho}} = 0.$$

These last two facts used in conjunction with (5.26) and (5.32) gives us that

$$\lim_{n\to\infty} (\lambda_{j_o+j_1} - \lambda_{j_o}) \|W_{n3}\|^2_{\tilde{L}^2_{\varrho}} = \lim_{n\to\infty} < f(\cdot,u_n), W_{n3} >_{\varrho}^{\tilde{}} / \|u_n\|_{\tilde{L}^2_{\varrho}}$$

and hence from (5.31) that

$$(\lambda_{j_o+j_1} - \lambda_{j_o}) \|W_3\|^2_{\tilde{L}^2_{\varrho}} = \lim_{n\to\infty} < f(\cdot,u_n), W_{n3} >_{\varrho}^{\tilde{}} / \|u_n\|_{\tilde{L}^2_{\varrho}}. \qquad (5.32')$$

Proceeding in a similar manner and using the fact that

$$\mathcal{L}^{\tilde{}}(u_n, W_{n2}) = \lambda_{j_o} < u_n, W_{n2} >_{\varrho}^{\tilde{}},$$

enables us to conclude from (5.32) that

$$0 = \lim_{n\to\infty} < f(\cdot,u_n), W_{n2} >_{\varrho}^{\tilde{}} / \|u_n\|_{\tilde{L}^2_{\varrho}}. \qquad (5.33)$$

Next, we obtain from (f-2) that

$$|f(x,u_n)| \leq 2\gamma \,|\, u_n(x,t)\,| + f_o(x) \quad \text{a.e. in } \tilde{\Omega} \qquad (5.34)$$

where $f_o \in L^2_{\varrho}(\Omega)$, $2\gamma = \lambda_{j_o+j_1} - \lambda_{j_o}$, and $n \geq n_1$. Hence it follows from (5.4) that

$$\exists K_8 \text{ and } n_2 \text{ s.t.} \left\| \frac{f(\cdot,u_n)}{\|u_n\|_{\tilde{L}^2_{\varrho}}} \right\|_{\tilde{L}^2_{\varrho}} \leq K_8 \quad \forall n \geq n_2 \qquad (5.35)$$

where $n_2 \geq n_1$. But then we can invoke the Banach-Saks theorem and other well-known facts about Hilbert spaces (see [BJS, p. 181]) to obtain the following (where we use a full sequence for ease of notation):

$\exists \tilde{F} \in \tilde{L}^2_\varrho$ such that

$$(i) \lim_{n\to\infty} < f(\cdot, u_n) \|u_n\|^{-1}_{\tilde{L}^2_\varrho}, v >_{\tilde{\varrho}} = < \tilde{F}, v >_{\tilde{\varrho}} \quad \forall v \in \tilde{L}^2_\varrho;$$

$$(ii) \lim_{n\to\infty} \left\| n^{-1} \sum_{k=n_2}^{n_2+n} f(\cdot, u_k) \|u_k\|^{-1}_{\tilde{L}^2_\varrho} - \tilde{F} \right\|_{\tilde{L}^2_\varrho} = 0; \tag{5.36}$$

(iii) for a certain subsequence $\{n_j\}_{j=1}^\infty$,

$$\lim_{j\to\infty} n_j^{-1} \sum_{k=n_2}^{n_2+n_j} f(x, u_k) \|u_k\|^{-1}_{\tilde{L}^2_\varrho} = \tilde{F}(x,t) \text{ a.e. in } \tilde{\Omega}.$$

From (5.31), (5.32′), and (5.35), we next see that

$$(\lambda_{j_o+j_1} - \lambda_{j_o}) \|W_3\|^2_{\tilde{L}^2_\varrho} = \lim_{n\to\infty} < f(\cdot, u_n) \|u_n\|^{-1}_{\tilde{L}^2_\varrho}, W_3 >_{\tilde{\varrho}}$$

But then it follows from (5.36)(ii) that

$$< \tilde{F}, W_3 >_{\tilde{\varrho}} = (\lambda_{j_o+j_1} - \lambda_{j_o}) \|W_3\|^2_{\tilde{L}^2_\varrho}. \tag{5.37}$$

In a similar manner, we obtain from (5.33) that

$$< \tilde{F}, W_2 >_{\tilde{\varrho}} = 0. \tag{5.38}$$

Proceeding along these same lines, we next observe from (5.34) that

$$\begin{aligned} \left| n^{-1} \sum_{k=n_2}^{n_2+n} f(x, u_k) \|u_k\|^{-1}_{\tilde{L}^2_\varrho} \right| \leq\ & 2\gamma n^{-1} \sum_{k=n_2}^{n_2+n} u_k(x,t) \|u_k\|^{-1}_{\tilde{L}^2_\varrho} \\ & + n^{-1} |f_o(x)| \sum_{k=n_2}^{n_2+n} \|u_k\|^{-1}_{\tilde{L}^2_\varrho} \end{aligned}$$

for a.e. x$\in \Omega$. Therefore, from (5.24)

$$\begin{aligned} \left| n^{-1} \sum_{k=n_2}^{n_2+n} f(x, u_k) \|u_k\|^{-1}_{\tilde{L}^2_\varrho} \right| \leq\ & 2\gamma n^{-1} \sum_{k=n_2}^{n_2+n} W_k(x,t) \\ & + n^{-1} |f_o(x)| \sum_{k=n_2}^{n_2+n} \|u_k\|^{-1}_{\tilde{L}^2_\varrho} \end{aligned}$$

for a.e. x$\in \Omega$, and we conclude from (5.27)(ii) and (5.36)(iii) that

$$\tilde{F}(x,t) = 0 \text{ a.e. in } \tilde{\Omega}_o = \Omega_o \times T \tag{5.39}$$

where $\Omega_o = \{x \in \Omega : W(x) = 0\}$.

Using $(f-2)$ once again, next we observe that

$$-\frac{f_o(x)}{|s|} \leq \frac{f(x,s)}{s} \leq 2\gamma + \frac{f_o(x)}{|s|}$$

for $s \neq 0$ and a.e. x$\in \Omega$. As a consequence, we see from $(1, 2.11)$ that

$$0 \leq \mathcal{F}_+(x) \leq \mathcal{F}^+(x) \leq 2\gamma \text{ and } 0 \leq \mathcal{F}_-(x) \leq \mathcal{F}^-(x) \leq 2\gamma \tag{5.40}$$

for a.e. x$\in \Omega$ where $2\gamma = (\lambda_{j_o+j_1} - \lambda_{j_o})$. Also, we define

$$\tilde{\Omega}_{1+} = \Omega_{1+} \times T \text{ and } \tilde{\Omega}_{1-} = \Omega_{1-} \times T$$

where

$$\Omega_{1+} = \{x \in \Omega : W(x) > 0\} \text{ and } \Omega_{1-} = \{x \in \Omega : W(x) < 0\}. \tag{5.41}$$

Let $(x_o, t_o) \in \tilde{\Omega}_{1+}$ be such that $\tilde{F}(x_o, t_o)$ is finite, the limits defined in $(5.27)\,(ii)$ and $(5.36)\,(iii)$ hold, and x_o is a value such that the inequalities in (5.40) hold. Then given $\varepsilon > 0$, it follows from the definition of $\mathcal{F}^+(x_o)$ that

$$f(x_o, s) \leq \mathcal{F}^+(x_o)\, s + \varepsilon s \quad for\ s \geq s^*.$$

Now $u_n(x_o, t_o) = \|u_n\|_{\tilde{L}^2_\varrho} W_n(x_o, t_o)$. Therefore from (5.4),

$$\lim_{n \to \infty} u_n(x_o, t_o) = \infty.$$

Hence, it follows from $(5.27)\,(ii)$, $(5.36)\,(iii)$, and this last inequality above that

$$\tilde{F}(x_o, t_o) \leq \mathcal{F}^+(x_o)\, W(x_o).$$

In a similar manner, it follows that

$$\mathcal{F}_+(x_o)\, W(x_o) \leq \tilde{F}(x_o, t_o).$$

Also, we see that a similar situation prevails for $\tilde{\Omega}_{1-}$. We conclude that

$$(i)\ \mathcal{F}_+(x)\, W(x) \leq \tilde{F}(x,t) \leq \mathcal{F}^+(x)\, W(x) \quad \text{a.e. in } \tilde{\Omega}_{1+}, \tag{5.42}$$

$$(ii)\ \mathcal{F}^{-}(x)W(x) \leq \tilde{F}(x,t) \leq \mathcal{F}_{-}(x)W(x) \quad \text{a.e. in } \tilde{\Omega}_{1-}.$$

Next, observing from (5.39) and (5.41)that

$$\tilde{\Omega} = \tilde{\Omega}_o \cup \tilde{\Omega}_{1+} \cup \tilde{\Omega}_{1-}$$

we define the function

$$\begin{aligned} F(x,t) &= 0 && (x,t) \in \tilde{\Omega}_o \\ &= \tilde{F}(x,t)/W(x) && (x,t) \in \tilde{\Omega}_{1+} \cup \tilde{\Omega}_{1-}. \end{aligned}$$

Hence, from (5.39)

$$\tilde{F}(x,t) = F(x,t)W(x) \qquad \text{a.e. in } \tilde{\Omega}. \tag{5.43}$$

Also from (5.40) and (5.42), we obtain

$$0 \leq F(x,t) \leq 2\gamma \qquad \text{a.e. in } \tilde{\Omega}. \tag{5.44}$$

Continuing in this vein, we see from (5.38) and (5.43) that

$$< FW, W_2 >_{\tilde{\varrho}} = 0,$$

and therefore from (5.30) that

$$< FW_2, W_2 >_{\tilde{\varrho}} = - < FW_3, W_2 >_{\tilde{\varrho}}.$$

Likewise from (5.30)and (5.37) we obtain

$$< (\lambda_{j_o+j_1} - \lambda_{j_o} - F)W_3, W_3 >_{\tilde{\varrho}} = < FW_2, W_3 >_{\tilde{\varrho}}.$$

Adding these last two equalities gives us that

$$< (\lambda_{j_o+j_1} - \lambda_{j_o} - F)W_3, W_3 >_{\tilde{\varrho}} + < FW_2, W_2 >_{\tilde{\varrho}} = 0.$$

But $2\gamma = \lambda_{j_o+j_1} - \lambda_{j_o}$; so from (5.44) both terms on the left-hand side of this last stated equation are nonnegative. Consequently, we have

$$(i) < (\lambda_{j_o+j_1} - \lambda_{j_o} - F)W_3, W_3 >_{\tilde{\varrho}} = 0, \tag{5.45}$$

$$(ii) < FW_2, W_2 >_{\tilde{\varrho}} = 0.$$

Next, we set

$$\tilde{\Omega}_2 = \Omega_2 \times T \text{ and } \tilde{\Omega}_3 = \Omega_3 \times T$$

where

$$\Omega_i = \{x \in \Omega : W_i(x) \neq 0\} \quad i = 2, 3. \tag{5.46}$$

From (5.44) and (5.45), we see that $F(x,t) = 0$ a.e. in $\tilde{\Omega}_2$ and $F(x,t) = \lambda_{j_o+j_1} - \lambda_{j_o}$ a.e. in $\tilde{\Omega}_3$. Consequently, $\tilde{\Omega}_2 \cap \tilde{\Omega}_3$ is a set of Lebesgue measure zero. But by (5.30), both W_2 and W_3 are continuous functions in Ω. Therefore both Ω_2 and Ω_3 are open sets, and we have that

$$\Omega_2 \text{ and } \Omega_3 \text{ are disjoint sets.} \tag{5.47}$$

Since $W = W_2 + W_3$, we see from (5.46) and (5.47)

$$W = W_2 \text{ on } \Omega_2, \text{ and } W = W_3 \text{ on } \Omega_3. \tag{5.48}$$

Next, we set

$$\tilde{\Omega}_{i+} = \Omega_{i+} \times T \text{ and } \tilde{\Omega}_{i-} = \Omega_{i-} \times T \quad \text{i=2,3} \tag{5.49}$$

where

$$(i)\, \Omega_{i+} = \{x \in \Omega : W_i(x) > 0\} \quad i = 2, 3,$$

$$(i)\, \Omega_{i-} = \{x \in \Omega : W_i(x) < 0\} \quad i = 2, 3.$$

Now from (5.30) we see that $W_2(x)$ is a λ_{j_o}−eigenfunction for L. Hence, it follows from (1,2.13) that if W_2 is nontrivial,

$$0 < \int_{\Omega_{2+}} \mathcal{F}_+(x)\, W_2^2(x)\, \varrho(x)\, dx + \int_{\Omega_{2-}} \mathcal{F}_-(x)\, W_2^2(x)\, \varrho(x)\, dx,$$

and therefore from (5.48)

$$0 < \int_{\Omega_{2+}} \mathcal{F}_+(x)\, W(x)\, W_2(x)\, \varrho(x)\, dx$$

$$+ \int_{\Omega_{2-}} \mathcal{F}_-(x)\, W(x)\, W_2(x)\, \varrho(x)\, dx$$

But then from (5.49), we have that

$$0 < \int_{\tilde{\Omega}_{2+}} \mathcal{F}_+(x)\, W(x)\, W_2(x)\, \varrho(x)\, dxdt \tag{5.50}$$

$$+ \int_{\tilde{\Omega}_{2-}} \mathcal{F}_-(x)\, W(x)\, W_2(x)\, \varrho(x)\, dxdt$$

We next observe from (5.48), the lines below (5.49), and (5.41) that if $(x,t) \in \tilde{\Omega}_{2+}$, then $(x,t) \in \tilde{\Omega}_{1+}$. Therefore from (5.42)(i)

$$\mathcal{F}_+(x)\, W(x)\, W_2(x) \leq \tilde{F}(x,t)\, W_2(x) \qquad \text{a.e. on } \tilde{\Omega}_{2+}.$$

Likewise from (5.42)(ii), we see that

$$\mathcal{F}_-(x)\, W(x)\, W_2(x) \leq \tilde{F}(x,t)\, W_2(x) \qquad \text{a.e. on } \tilde{\Omega}_{2-}.$$

Using these last two inequalities in conjunction with the inequality in (5.50) then gives us that

$$0 < \int_{\tilde{\Omega}_{2+}} \tilde{F}(x,t)\, W_2(x)\, \varrho(x)\, dxdt + \int_{\tilde{\Omega}_{2-}} \tilde{F}(x,t)\, W_2(x)\, \varrho(x)\, dxdt$$

Consequently,

$$< \tilde{F}, W_2 >_{\tilde{\varrho}} > 0.$$

This inequality is based on a line of reasoning starting with the assumption that W_2 is a nontrivial λ_{j_o}-eigenfunction of L. However by (5.38)

$$< \tilde{F}, W_2 >_{\tilde{\varrho}} = 0.$$

Consequently, we must conclude that W_2 is indeed trivial, i.e.,

$$W_2(x) = 0 \quad \text{for all } x \in \Omega.$$

As a further consequence of this last fact, we have from (5.30)

$$W(x) = W_3(x) \quad \text{for all } x \in \Omega. \tag{5.51}$$

Now as we have observed earlier, W_3 is a $\lambda_{j_o+j_1}-$ eigenfunction of L. Hence, from (5.51) so is W. If furthermore W is a nontrivial function, then it follows from (1,2.12) that

$$\begin{aligned} 2\gamma \int_\Omega |\, W(x)\,|^2\, \varrho(x)\; dx > & \int_{\Omega_{1+}} \mathcal{F}^+(x)\, W^2(x)\, \varrho(x)\, dx \\ & + \int_{\Omega_{1-}} \mathcal{F}^-(x)\, W^2(x)\, \varrho(x)\, dx \end{aligned}$$

where $2\gamma = \lambda_{j_o+j_1}$. As a consequence, we see from the line above (5.41)

$$\begin{aligned} 2\gamma\, \|W\|^2_{\tilde{L}^2_\varrho} \quad > \quad & \int_{\tilde{\Omega}_{1+}} \mathcal{F}^+(x)\, W^2(x)\, \varrho(x)\, dxdt \\ & + \int_{\tilde{\Omega}_{1-}} \mathcal{F}^-(x)\, W^2(x)\, \varrho(x)\, dxdt. \end{aligned} \tag{5.52}$$

Next, we obtain from (5.42)(i)

$$\tilde{F}(x,t)\,W(x) \leq \mathcal{F}^{+}(x)\,W^{2}(x) \quad \text{for a.e. } (x,t) \in \tilde{\Omega}_{1+}$$

and from (5.42)(ii)

$$\tilde{F}(x,t)\,W(x) \leq \mathcal{F}^{-}(x)\,W^{2}(x) \quad \text{for a.e. } (x,t) \in \tilde{\Omega}_{1-}.$$

Therefore, we have from (5.52) that

$$\begin{aligned} 2\gamma \|W\|^{2}_{\tilde{L}^{2}_{\varrho}} &> \int_{\tilde{\Omega}_{1+}} \tilde{F}(x,t)\,W(x)\,\varrho(x)\,dxdt \\ &+ \int_{\tilde{\Omega}_{1-}} \tilde{F}(x,t)\,W(x)\,\varrho(x)\,dxdt. \end{aligned}$$

But the right-hand side of the above inequality is equal to $<\tilde{F},W>_{\tilde{\varrho}}$. Hence we see that the following fact obtains:

$$<\tilde{F},W>_{\tilde{\varrho}} < (\lambda_{j_o+j_1} - \lambda_{j_o})\,\|W\|^{2}_{\tilde{L}^{2}_{\varrho}}. \tag{5.53}$$

Now this last inequality was obtained on the assumption that W(x) was a nontrivial function. However, from (5.51), W(x) $=$ W_3(x) for all x in Ω. Consequently, it follows from (5.37) that

$$<\tilde{F},W>_{\tilde{\varrho}} = (\lambda_{j_o+j_1} - \lambda_{j_o})\,\|W\|^{2}_{\tilde{L}^{2}_{\varrho}}.$$

But this last equality is a direct contradiction of the inequality in (5.53), and we therefore conclude that

$$W(x) = 0 \quad \forall x \in \Omega.$$

Next, we take $W = 0$ and insert it in the expression (5.27)(i) and obtain

$$\lim_{n\to\infty} \|W_n\|_{\tilde{L}^{2}_{\varrho}} = 0.$$

On the other hand, from (5.24) we see that

$$\lim_{n\to\infty} \|W_n\|_{\tilde{L}^{2}_{\varrho}} = 1.$$

Hence, 1=0, a manifest contradiction! All of this was obtained on the basis that (5.3) was true. Therefore, we conclude that (5.3) is false, and (5.2) is indeed true.

With (5.2) and (5.1) at our disposal, we proceed with proof of our theorem along the lines given for the proof, at this point, of Theorem 1 in Chapter 1. In particular, we see from Lemma 3 and the observation that $\tilde{\mathrm{H}}$ is a separable Hilbert space,that the following facts obtain for the sequence $\{u_n\}_{n=n_1}^{\infty}$ which satisfies (5.1) and (5.2) above:

$\exists\ u^* \in \tilde{H}$ and a subsequence (which for ease of notation we take to be the full sequence) such that

$$\lim_{n\to\infty} \|u_n - u^*\|_{\tilde{L}^2_\varrho} = 0, \tag{5.54}$$

$$\exists w \in \tilde{L}^2_\varrho \ s.t.\ |u_n(x,t)| \le w(x,t) \tag{5.55}$$

$$\text{for a.e. } (x,t) \in \tilde{\Omega}, \quad n = n_1, n_1+1, \ldots,$$

$$\lim_{n\to\infty} u_n(x,t) = u^*(x,t) \text{ for } a.e.\ (x,t) \in \tilde{\Omega}, \tag{5.56}$$

$$\lim_{n\to\infty} < D_i u_n, v >_{\tilde{p_i}} = < D_i u^*, v >_{\tilde{p_i}} \forall v \in \tilde{L}^2_{p_i} \text{ and } i = 1, ..., N, \tag{5.57}$$

$$\lim_{n\to\infty} < D_t u_n, v >_{\tilde{\varrho}} = < D_t u^*, v >_{\tilde{\varrho}} \forall v \in \tilde{L}^2_\varrho. \tag{5.58}$$

Next, we propose to show the following: there exists a subsequence $\{u_{n_j}\}_{j=1}^{\infty}$such that

$$\lim_{j\to\infty} D_i u_{n_j}(x,t) = D_i u^*(x,t) \quad \text{for } a.e.\ (x,t) \in \tilde{\Omega} \tag{5.59}$$

for $i = 1, \ldots, N$. To establish (5.59), it is sufficient to establish the following two facts:

(1) There exists a subsequence $\{u_{n_j}\}_{j=1}^{\infty}$ such that

$$\lim_{j\to\infty} \sum_{i=1}^{N} p_i^{\frac{1}{2}}(x) \{ [\, A_i(x, u_{n_j}(x,t), Du_{n_j}(x,t)) - A_i(x, u_{n_j}(x,t), \tag{5.60}$$

$$Du^*(x,t))\,]\, [D_i u_{n_j}(x,t) - D_i u^*(x,t)]\,\} = 0 \text{ for } a.e.\ (x,t) \in \tilde{\Omega}.$$

(2) With $\{u_{n_j}\}_{j=1}^{\infty}$ designating the same subsequence as in (5.60),

$$\{|\, D_i u_{n_j}(x,t)\,|\}_{j=1}^{\infty} \text{ is pointwise bounded for } a.e.\ (x,t) \in \tilde{\Omega} \tag{5.61}$$

for i=1,...,N.

The proof that (5.60) and (5.61) together imply (5.59) is very similar to the proof of the analogous situation given in Chapter 1 where we showed

(1,5.26) and (1,5.27) together imply (1,5.25). So we leave the details of the above implication, namely that (5.60) and (5.61) together imply (5.59), to the reader.

It remains to show that (5.60) and (5.61) hold.

To establish (5.60), we observe from (5.55)-(5.57), and (5.3) that the analogues of (1,5.31), (1,5.32), and (1,5.33) hold. Hence, we have

$$\lim_{n\to\infty}\int_{\tilde{\Omega}}\sum_{i=1}^{N}A_i\left(x,u_n,Du^*\right)\left[D_iu_n-D_iu^*\right]p_i^{\frac{1}{2}}=0. \tag{5.62}$$

Likewise, we have from (1.1), (1.7)(ii),and (5.7) that

$$\lim_{n\to\infty}\int_{\tilde{\Omega}}B_0\left(x,u_n,Du_n\right)u^*\left(u_n-u^*\right)q=0. \tag{5.63}$$

Now if we can show that

$$\lim_{n\to\infty}\mathcal{Q}^{\tilde{}}\left(u_n,u_n-u^*\right)=0, \tag{5.64}$$

then it will follow from (1.9), (5.62), and (5.63) that

$$\lim_{n\to\infty}\int_{\tilde{\Omega}}\{\sum_{i=1}^{N}[A_i(x,u_n,Du_n)-A_i(x,u_n,Du^*)]\left[D_iu_n-D_iu^*\right]p_i^{\frac{1}{2}} \tag{5.65}$$

$$+B_0(x,u_n,Du_n)\left(u_n-u^*\right)\left(u_n-u^*\right)q\}=0$$

This last fact, in turn, implies (5.60). For from (Q-4) and (1.7)(ii), we have that the integrand in (5.65) is nonnegative almost everywhere in Ω. Hence the integrand converges in $L^1(\tilde{\Omega})$ to zero. But then by $[Rud, p.70]$, a subsequence of the integrand converges almost everywhere in $\tilde{\Omega}$ to zero. By (1.7)(ii), and (5.56) we have that $B_0(x,u_n,Du_n)\left|u_n\left(x,t\right)-u^*\left(x,t\right)\right|^2$ converges to zero almost everywhere in $\tilde{\Omega}$. We conclude that (5.60) is indeed true. So to establish (5.60), it remains to establish (5.64).

To show that (5.64) is indeed true, we observe that $u^*\in\tilde{H}$ and from (4.8) that

$$\tau_n\left(u^*\right)\in S_n^{\ddagger}.$$

Also from Lemma 1, we have that

$$\lim_{n\to\infty}\left\|\tau_n\left(u^*\right)-u^*\right\|_{\tilde{H}}=0. \tag{5.66}$$

Now

$$\mathcal{Q}^{\tilde{}}(u_n, \tau_n(u^*) - u^*) = \sum_{i=1}^{N} \int_{\Omega} A_i(x, u_n, Du_n)\, D_i(\tau_n(u^*) - u^*) p_i^{\frac{1}{2}}$$

$$+ < B_0(\cdot, u_n, Du_n)\, u_n, \tau_n(u^*) - u^* >_{\tilde{q}} .$$

and it follows from (Q-2), (Q-5), (5.2), and (5.66) that

$$\lim_{n \to \infty} \mathcal{Q}^{\tilde{}}(u_n, \tau_n(u^*) - u^*) = 0.$$

Therefore (5.64) will follow if we show

$$\lim_{n \to \infty} \mathcal{Q}^{\tilde{}}(u_n, u_n - \tau_n(u^*)) = 0. \tag{5.67}$$

Now from (5.1), we see that

$$< D_t u_n, u_n - \tau_n(u^*) >_{\tilde{\varrho}} + \mathcal{Q}^{\tilde{}}(u_n, u_n - \tau_n(u^*))$$

$$= (\lambda_{j_o} + \gamma n^{-1}) < u_n, u_n - \tau_n(u^*) >_{\tilde{\varrho}} \tag{5.68}$$

$$+(1 - n^{-1}) < f(\cdot, u_n) + g(\cdot, \cdot, u_n), u_n - \tau_n(u^*) >_{\tilde{\varrho}}$$

But from (5.2), (5.54), and (5.66), we see that

$$\lim_{n \to \infty} < D_t u_n, u_n - \tau_n(u^*) >_{\tilde{\varrho}} = 0,$$

and that

$$\lim_{n \to \infty} < u_n, u_n - \tau_n(u^*) >_{\tilde{\varrho}} = 0.$$

Consequently, it follows from (5.68) that (5.67) will be established once we show

$$\lim_{n \to \infty} < f(\cdot, u_n) + g(\cdot, \cdot, u_n), u_n - \tau_n(u^*) >_{\tilde{\varrho}} = 0. \tag{5.69}$$

From (f-2) and (5.34), we see that

$$|f(x, u_n)| \le 2\gamma \,|\, u_n(x,t) \,| + f_o(x) \quad \text{a.e. in } \tilde{\Omega} \tag{5.70}$$

where $f_o \in L^2_{\varrho}(\Omega), 2\gamma = \lambda_{j_o + j_1} - \lambda_{j_o}$, and $n \ge n_1$. Also, from (g-2), we have that

$$|g(x, t, u_n)| \le |\, u_n(x,t) \,| + g_1(x,t) \quad \text{a.e. in } \tilde{\Omega} \tag{5.71}$$

where $g_1 \in L^2_{\varrho}\left(\tilde{\Omega}\right) = \tilde{L}^2_{\varrho}$.

These last two facts in conjunction with (5.54), (5.66), and(5.2) show that (5.69) is indeed true. Hence, the same goes for (5.67) and (5.64), and we conclude that (5.60) is completely established.

Observing that

$$|Du(x,t)| = \left[|D_1 u(x,t)|^2 + \ldots + |D_n u(x,t)|^2\right]^{\frac{1}{2}},$$

we see that the proof of (5.61) using (5.60) follows along the same lines that we used to obtain (1,5.27) using (1,5.26). Proper modifications have to be made for the fact that we are now working in $\tilde{\Omega}$ rather than Ω, but the ideas used are essentially the same. We leave the details of the proof to the reader (See (6.51)-(6.55) below where we do put in the details of an analogous situation involved in the proof of Theorem 2.) and obtain that (5.61) is indeed true. As we have remarked earlier (5.60) and (5.61) together imply (5.59). Consequently, (5.59) is also established.

We now proceed with the proof of the theorem using (5.1), (5.2), and (5.54)-(5.59). Let $v \in \tilde{H}$ and let $\tau_J(v)$ be defined by (4.8). Then $\tau_J(v) \in S_J^{\ddagger}$ and from (5.1) with $J \geq n_1$, we have that

$$< D_t u_n, \tau_J(v) >_{\varrho}^{\tilde{}} + \mathcal{Q}^{\tilde{}}(u_n, \tau_J(v)) = (\lambda_{j_o} + \gamma n^{-1}) < u_n, \tau_J(v) >_{\varrho}^{\tilde{}}$$
$$+(1-n^{-1}) < f(\cdot, u_n) + g(\cdot,\cdot,u_n), \tau_J(v) >_{\varrho}^{\tilde{}}. \tag{5.72}$$

for $n \geq J$.

From (5.2) and (Q-2) we have that the sequence

$$\left\{\int_{\tilde{\Omega}} \left|A_i(x, u_{n_j}, Du_{n_j}) - A_i(x, u^*, Du^*)\right|^2\right\}_{j=1}^{\infty}$$

is a uniformly bounded sequence for i=1,...,N Also, it follows from (5.56) and (5.59) that

$$\lim_{j\to\infty} \left|A_i(x, u_{n_j}, Du_{n_j}) - A_i(x, u^*, Du^*)\right| / p_i^{\frac{1}{2}} = 0 \text{ a.e. in } \tilde{\Omega}.$$

Now $D_i\tau_J(v) \in \tilde{L}^2_{p_i}$. Consequently, we obtain from Schwarz's inequality and from Egoroff's theorem (see [Rud, p. 75]) applied to the positive measure

$$\nu_i(E) = \int_E p_i(x)\, dxdt \quad \text{for } E \subset \tilde{\Omega}$$

for $i = 1, \ldots, N$, where E is a Lebesgue measurable set that,

$$\lim_{j\to\infty} \int_{\tilde{\Omega}} \left[A_i(x, u_{n_j}, Du_{n_j}) - A_i(x, u^*, Du^*)\right] D_i\tau_J(v)\ p_i^{\frac{1}{2}} = 0.$$

We conclude from all this and (1.9) that

$$\lim_{n\to\infty} \mathcal{Q}^{\tilde{}}(u_n, \tau_J(v)) = \mathcal{Q}^{\tilde{}}(u^*, \tau_J(v))$$

Next, from (5.70) and (5.55), we have that

$$|f(x, u_n)| \leq 2\gamma \,|\, w(x,t) \,|\, + f_o(x) \quad \text{a.e. in } \tilde{\Omega},$$

where $w \in \tilde{L}^2_\varrho$ and $f_o \in L^2_\varrho$. Likewise, from (5.71), we have that

$$|g(x, t, u_n)| \leq |\, w(x,t) \,|\, + g_1(x,t) \quad \text{a.e. in } \tilde{\Omega},$$

where $g_1 \in \tilde{L}^2_\varrho$. So we conclude from the Lebesgue dominated convergence theorem in conjunction with (5.56) that

$$\lim_{n\to\infty} < f(\cdot, u_n) + g(\cdot, \cdot, u_n), \tau_J(v) >_{\varrho}^{\tilde{}} = < f(\cdot, u^*) + g(\cdot, \cdot, u^*), \tau_J(v) >_{\varrho}^{\tilde{}} .$$

From these last two limits joined with those in (5.54) and (5.58), we see in turn from (5.72) that

$$< D_t u^*, \tau_J(v) >_{\varrho}^{\tilde{}} + \mathcal{Q}^{\tilde{}}(u^*, \tau_J(v)) = \tag{5.73}$$

$$\lambda_{j_o} < u^*, \tau_J(v) >_{\varrho}^{\tilde{}} + < f(\cdot, u^*) + g(\cdot, \cdot, u^*), \tau_J(v) >_{\varrho}^{\tilde{}} .$$

But, $\tau_J(v) \to v$ in $\tilde{H}$ as $J \to \infty$; so taking the limit on both on both sides of the inequality in (5.73), we have that

$$< D_t u^*, v >_{\varrho}^{\tilde{}} + \mathcal{Q}^{\tilde{}}(u^*, v) = \lambda_{j_o} < u^*, v >_{\varrho}^{\tilde{}}$$

$$+ < f(\cdot, u^*) + g(\cdot, \cdot, u^*), v >_{\varrho}^{\tilde{}}$$

for all $v \in \tilde{H}$, and the proof of Theorem 1 is complete.

2.6 Proof of Theorem 2

To prove Theorem 6, we invoke Lemma 5 and have a sequence

$$\{(u_{1n}, \ldots, u_{N_1+N_2+N_3 n})\}_{n=n_o^*+1}^{\infty} \text{ with } u_{in} \in S_n^{\ddagger} \text{ (defined in (4.23))}$$

such that for n>n_0^* (defined in (4.39)) the following obtains:

$$< D_t u_{in}, v >_{\varrho}^{\tilde{}} + \mu_i \mathcal{Q}_i^{\tilde{}}(u_{in}, v) = \tag{6.1}$$

$$\mu_i(\lambda_{j_o(i)} + \frac{1}{n}) < u_{in}, v >_{\tilde{\varrho}} + < f_i(\cdot,\cdot,u_{in}) + g_i(\cdot,\cdot,u_{1n},\ldots,u_{N_1+N_2+N_3 n})$$

$$-h_i(\cdot,\cdot), v >_{\tilde{\varrho}} \quad \forall v \in S_n^{\ddagger}, \qquad i = 1,\ldots,N_1,$$

$$\mu_i(\lambda_{j_o(i)} + \varepsilon_i) < u_{in}, v >_{\tilde{\varrho}} + < f_i(\cdot,\cdot,u_{in}) + g_i(\cdot,\cdot,u_{1n},\ldots,u_{N_1+N_2+N_3 n})$$

$$-h_i(\cdot,\cdot), v >_{\tilde{\varrho}} \quad \forall v \in S_n^{\ddagger}, \ i = N_1+1,\ldots,N_1+N_2,$$

$$\mu_i(\lambda_1 - \varepsilon_i) < u_{in}, v >_{\tilde{\varrho}} + < f_i(\cdot,\cdot,u_{in}) + g_i(\cdot,\cdot,u_{1n},\ldots,u_{N_1+N_2+N_3 n})$$

$$-h_i(\cdot,\cdot), v >_{\tilde{\varrho}} \quad \forall v \in S_n^{\ddagger}, \ i = N_1+N_2+1,\ldots,N_1+N_2+N_3.$$

With $\|v\|_{\mathcal{L}^{\tilde{}}}$ defined in (2.4), we claim there is a constant K_1 such that

$$\|u_{in}\|_{\tilde{L}^2_\varrho} + \|u_{in}\|_{\mathcal{L}^{\tilde{}}} \leq K_1 \qquad \forall n > n_0^* \tag{6.2}$$

for i=1,..., N_1 +N_2+ N_3 where $\|u_{in}\|^2_{\tilde{L}^2_\varrho} =< u_{in}, u_{in} >_{\tilde{\varrho}}$.

We shall establish (6.2) separately for the three cases represented by Case I when i=1,..., N_1, Case II when i=N_1+1,..., N_1 +N_2, and Case III when i=N_1 +N_2+1,..., N_1 +N_2+ N_3.

For ease of notation, in the proof for Case I, we shall establish (6.2) for i=1. A similar proof will prevail for i=2,..., N_1. Also, we shall assume $j_o(1)$ >1. For $j_o(1)$ =1, a similar proof will work. Suppose then that (6.2) is false for the sequence

$$\left\{ \|u_{1n}\|_{\tilde{L}^2_\varrho} + \|u_{1n}\|_{\mathcal{L}^{\tilde{}}} \right\}_{n=n_o^*+1}^{\infty}$$

Then without loss in generality, we can suppose that

$$\lim_{n\to\infty} [\|u_{1n}\|_{\tilde{L}^2_\varrho} + \|u_{1n}\|_{\mathcal{L}^{\tilde{}}}] = \infty. \tag{6.3}$$

We propose to show that (6.3) leads to a contradiction of the inequality in (2.8) when i=1. To accomplish this, we put u_{1n} in place of v in (6.1) to obtain

$$\mu_1 \tilde{Q_1}(u_{1n}, u_{1n}) - \mu_1(\lambda_{j_o(1)} + \frac{1}{n}) < u_{1n}, u_{1n} >_{\tilde{\varrho}} = \tag{6.4}$$

$$< f_1(\cdot,\cdot,u_{1n}) + g_1(\cdot,\cdot,u_{1n},\ldots,u_{N_1+N_2+N_3 n}) - h_1(\cdot,\cdot), u_{1n} >_{\tilde{\varrho}}$$

where we have used the fact (see (4.25)) that

$$< u_{1n}, D_t u_{1n} >_{\tilde{\varrho}} = 0. \tag{6.5}$$

We next see that (6.3) and (6.4) imply that there exists a positive constant K_2 such that

$$\|u_{1n}\|_{\mathcal{L}^{-}} \leq K_2 \|u_{1n}\|_{\tilde{L}^2_{\varrho}} \quad \forall n > n^*_o. \tag{6.6}$$

For suppose that (6.6) is false. Then there exists a subsequence (which we take to be the full sequence) $\{\eta_n\}_{n=n^*_o+1}^{\infty}$ with $\eta_n \to \infty$ such that

$$\|u_{1n}\|_{\mathcal{L}^{-}} \geq \eta_n \|u_{1n}\|_{\tilde{L}^2_{\varrho}} \quad \forall n > n^*_o.$$

But then in view of (6.3), $\|u_{1n}\|_{\mathcal{L}^{-}} \to \infty$. Hence we observe from (2.3)(ii), (2.4), and (6.3) that there exists a positive constant c^*_1 and an integer n^*_1 such that.

$$c^*_1 \|u_{1n}\|^2_{\mathcal{L}^{-}} \leq \mu_1 \mathcal{Q}^{-}_1(u_{1n}, u_{1n}) \quad \forall n > n^*_1.$$

As a consequence, we see from (f-10), (g-4), and this last inequality applied to the equation in (6.4) that first of all $\|u_{1n}\|_{\mathcal{L}^{-}} \to \infty$ implies that

$$\lim_{n\to\infty} \|u_{1n}\|_{\tilde{L}^2_{\varrho}} = \infty,$$

and secondly that there exists a positive constant K'_2 such that

$$\|u_{1n}\|_{\mathcal{L}^{-}} \leq K'_2 \|u_{1n}\|_{\tilde{L}^2_{\varrho}} \quad \forall n > n^*_1.$$

Therefore from the above, we have that

$$\eta_n \|u_{1n}\|_{\tilde{L}^2_{\varrho}} \leq K'_2 \|u_{1n}\|_{\tilde{L}^2_{\varrho}} \quad \forall n > n^*_1.$$

Consequently, the sequence $\{\eta_n\}_{n=n^*_o+1}^{\infty}$ is a bounded sequence, which contradicts the fact that $\eta_n \to \infty$. Hence (6.6) is indeed true

Next, we write

$$u_{1n} = u_{1n1} + u_{1n2} + u_{1n3} \text{ and } u^{\approx}_{1n} = -u_{1n1} - u_{1n2} + u_{1n3} \tag{6.7}$$

$$u_{1n2} = \sum_{j=j_o}^{j_o+j_1-1} \hat{u}^c_{1n}(j,0)\tilde{\varphi}^c_{j0} + \sum_{j=j_o}^{j_o+j_1-1}\sum_{k=1}^{n}[\hat{u}^c_{1n}(j,k)\tilde{\varphi}^c_{jk} + \hat{u}^s_{1n}(j,k)\tilde{\varphi}^s_{jk}].$$

where $j_o = j_o(1)$ and $j_1 = j_1(1)$ and

$$u_{1n1} = \sum_{j=1}^{j_o-1} \hat{u}^c_{1n}(j,0)\tilde{\varphi}^c_{j0} + \sum_{j=1}^{j_o-1}\sum_{k=1}^{n}[\hat{u}^c_{1n}(j,k)\tilde{\varphi}^c_{jk} + \hat{u}^s_{1n}(j,k)\tilde{\varphi}^s_{jk}],$$

$$u_{1n3} = \sum_{j=j_o+j_1}^{n} \hat{u}_{1n}^c(j,0)\tilde{\varphi}_{j0}^c + \sum_{j=j_o+j_1}^{n} \sum_{k=1}^{n} [\hat{u}_{1n}^c(j,k)\tilde{\varphi}_{jk}^c + \hat{u}_{1n}^s(j,k)\tilde{\varphi}_{jk}^s].$$

We claim the following:

$$(i) \lim_{n\to\infty} \|D_t u_{1n}\|_{\tilde{L}^2_\varrho} / \|u_{1n}\|_{\tilde{L}^2_\varrho} = 0; \tag{6.8}$$

$$(ii) \lim_{n\to\infty} [\|u_{1n1}\|^2_{\tilde{L}^2_\varrho} + \|u_{1n3}\|^2_{\tilde{\mathcal{L}}}]/ \|u_{1n}\|^2_{\tilde{L}^2_\varrho} = 0;$$

$$(iii) \lim_{n\to\infty} [\|u_{1n1}\|^2_{\tilde{H}} + \|u_{1n3}\|^2_{\tilde{H}}]/ \|u_{1n}\|^2_{\tilde{L}^2_\varrho} = 0;$$

To establish (i) in the above claim, we observe from (4.25)(i) that $D_t u_{1n} \in S_n^{\ddagger}$. Hence replacing v in (6.1) when i=1 by $D_t u_{1n}$, we obtain

$$\|D_t u_{1n}\|^2_{\tilde{L}^2_\varrho} = < f_1(\cdot,\cdot,u_{1n}) + g_1(\cdot,\cdot,u_{1n},\ldots,u_{N_1+N_2+N_3 n}) - h_1(\cdot,\cdot), D_t u_{1n} >_{\tilde{\varrho}} \tag{6.9}$$

where we have used (Q-6) with Remark 2 and (6.5). Using Schwarz's inequality in conjunction with (f-10) and (g-4), we then obtain from (6.9) that

$$\|D_t u_{1n}\|_{\tilde{L}^2_\varrho} \leq \|c\|_{\tilde{L}^2_\varrho} + \|c^*\|_{\tilde{L}^2_\varrho} + \|h_1\|_{\tilde{L}^2_\varrho}$$

Dividing both sides of this last inequality by $\|u_{1n}\|_{\tilde{L}^2_\varrho}$ and passing to the limit as n$\to \infty$, we conclude that claim (i) in (6.8) follows from (6.3) and (6.6).

As an immediate consequence of this last established claim, we see from (4.3), (4.4), and (6.6) that there exists a constant K_3 such that

$$\|u_{1n}\|_{\tilde{H}} \leq K_3 \|u_{1n}\|_{\tilde{L}^2_\varrho} \quad \text{for } n > n_o^*. \tag{6.10}$$

To establish claim (ii) in (6.8), we replace v in (6.1) when i=1 by $u_{1n}^{\approx}$, which is defined in (6.7), and obtain with $j_o = j_o(1)$ that

$$\mu_1 \tilde{\mathcal{Q}}_1(u_{1n}, u_{1n}^{\approx}) - \mu_1(\lambda_{j_o} + \frac{1}{n}) < u_{1n}, u_{1n}^{\approx} >_{\tilde{\varrho}} = \tag{6.11}$$

$$< f_1(\cdot,\cdot,u_{1n}) + g_1(\cdot,\cdot,u_{1n},\ldots,u_{N_1+N_2+N_3 n}) - h_1(\cdot,\cdot), u_{1n}^{\approx} >_{\tilde{\varrho}} .$$

In the above we have used the observation that

$$< D_t u_{in}, u_{1n}^{\approx} >_{\tilde{\varrho}} = 0,$$

a fact which follows easily from (4.25)(ii) and (6.7).

Next, we see from (4.6) and (6.7) that:

$$\|u_{1n}\|^2_{\mathcal{L}^{\tilde{}}} = \|u_{\tilde{1}n}^{\approx}\|^2_{\mathcal{L}^{\tilde{}}}\,;\;\; \|D_t u_{1n}\|^2_{\tilde{L}^2_\varrho} = \|D_t u_{\tilde{1}n}^{\approx}\|^2_{\tilde{L}^2_\varrho}\,;\;\; \|u_{1n}\|^2_{\tilde{L}^2_\varrho} = \|u_{\tilde{1}n}^{\approx}\|^2_{\tilde{L}^2_\varrho}\,. \tag{6.12}$$

Now by the hypothesis of the theorem, Q_1 is #-#$\tilde{H}$ related to L. Hence Q_1 is #$\tilde{H}$-related to L, and we obtain from (1.10), (4.4), (6.6), (6.10), and (6.12) that there exists a sequence $\left\{\varepsilon_n^{\#}\right\}_{n=n_1^*+1}^{\infty}$ with $\varepsilon_n^{\#} \to 0$ as n$\to \infty$ such that

$$\mu_1 \mathcal{Q}_1^{\tilde{}}\,(u_{1n}, u_{\tilde{1}n}^{\approx}) - \mu_1 \mathcal{L}^{\tilde{}}\,(u_{1n}, u_{\tilde{1}n}^{\approx}) = \varepsilon_n^{\#}\, \|u_{1n}\|^2_{\tilde{L}^2_\varrho}\,.$$

Consequently, we infer from (6.11), (6.12), (f-10), and (g-4) that

$$\left|\mu_1[\mathcal{L}^{\tilde{}}\,(u_{1n}, u_{\tilde{1}n}^{\approx}) - \lambda_{j_o} < u_{1n}, u_{\tilde{1}n}^{\approx} >_\varrho^{\tilde{}}]\right| \le \left(\varepsilon_n^{\#} + \frac{\mu_1}{n}\right) \|u_{1n}\|^2_{\tilde{L}^2_\varrho} \tag{6.13}$$

$$+ \|u_{1n}\|_{\tilde{L}^2_\varrho} \left[\|c\|_{\tilde{L}^2_\varrho} + \|c^*\|_{\tilde{L}^2_\varrho} + \|h_1\|_{\tilde{L}^2_\varrho}\right]$$

for n>n_o^*.

Now, the left-hand side of this last inequality from (4.6) and (6.7) is equal to

$$\mu_1 \{ \sum_{j=1}^{j_o-1} (\lambda_{j_o} - \lambda_j)\, |\hat{u}_{\tilde{1}n}^c(j,0)|^2 + \sum_{j=j_o+j_1}^{n} (\lambda_j - \lambda_{j_o})\, |\hat{u}_{\tilde{1}n}^c(j,0)|^2$$

$$+ \sum_{j=1}^{j_o-1} (\lambda_{j_o} - \lambda_j) \sum_{k=1}^{n} [|\hat{u}_{\tilde{1}n}^c(j,k)|^2 + |\hat{u}_{\tilde{1}n}^s(j,k)|^2]$$

$$+ \sum_{j=j_o+j_1}^{n} (\lambda_j - \lambda_{j_o}) \sum_{k=1}^{n} [|\hat{u}_{\tilde{1}n}^c(j,k)|^2 + |\hat{u}_{\tilde{1}n}^s(j,k)|^2]\, \}\,.$$

Also, we see that

$$\exists\, \delta > 0 \text{ s.t. } \lambda_j - \lambda_{j_o} \ge \delta\, \lambda_j \quad \forall\, j \ge j_o + j_1$$

$$\text{and s.t. } \lambda_{j_o} - \lambda_j \ge \delta \quad \text{for } 1 \le\; j \le j_o - 1.$$

So we conclude from (6.13)

$$\mu_1 \delta [< u_{1n1}, u_{1n1} >_\varrho^{\tilde{}} + \mathcal{L}^{\tilde{}}\,(u_{1n3}, u_{1n3}) \le \left(\varepsilon_n^{\#} + \frac{1}{n}\right) \|u_{1n}\|^2_{\tilde{L}^2_\varrho}$$

$$+ \|u_{1n}\|_{\tilde{L}^2_\varrho} \left[\|c\|_{\tilde{L}^2_\varrho} + \|c^*\|_{\tilde{L}^2_\varrho} + \|h_1\|_{\tilde{L}^2_\varrho} \right].$$

Dividing both sides of this last inequality by $\|u_{1n}\|^2_{\tilde{L}^2_\varrho}$,we see immediately from (6.3) and (6.6) that

$$\lim_{n\to\infty} [\|u_{1n1}\|^2_{\tilde{L}^2_\varrho} + \|u_{1n3}\|^2_{\tilde{\mathcal{L}}}] / \|u_{1n}\|^2_{\tilde{L}^2_\varrho} = 0,$$

and claim (ii) in (6.8) follows from this limit. Also, we note from (6.7) that

$$\|u_{1n1}\|^2_{\tilde{\mathcal{L}}} \leq \lambda_{j_o} \|u_{1n1}\|^2_{\tilde{L}^2_\varrho} \qquad \text{and} \tag{6.14}$$

$$\lambda_{j_o+j_1} < u_{1n3}, u_{1n3} >_{\tilde{\varrho}} \leq \tilde{\mathcal{L}} (u_{1n3}, u_{1n3}).$$

These facts in conjunction with claims (i) and (ii) in (6.8) establish the verity of claim (iii) in (6.8).

Proceeding with our proof that (6.3) leads to a contradiction of the inequality in (2.8) when i=1, we next define

$$W_{1n} = \frac{u_{1n}}{\|u_{1n}\|_{\tilde{L}^2_\varrho}} \text{ and } W_{1ni} = \frac{u_{1ni}}{\|u_{1n}\|_{\tilde{L}^2_\varrho}} \text{ for i=1,2,3,} \tag{6.15}$$

and observe from (6.7) that

$$\|u_{1n2}\|^2_{\tilde{\mathcal{L}}} = \lambda_{j_o} \|u_{1n2}\|^2_{\tilde{L}^2_\varrho}.$$

Therefore, $\|W_{1n2}\|^2_{\tilde{\mathcal{L}}} \leq \lambda_{j_o}$. As a consequence, we have from (i) and (iii) in (6.8) that there is a constant K_4 such that

$$\|W_{1n}\|^2_{\tilde{H}} \leq K_4 \qquad \forall n > n_o^*.$$

As a further consequence, we have from this last inequality in conjunction with Lemma 3, (iii) in (6.8), and (6.15) that the following facts obtain (where we are using a full sequence rather than a subsequence for ease of notation):

$$\exists\, W \in \tilde{H} \text{ such that} \tag{6.16}$$

$$(i) \lim_{n\to\infty} < W_{1n}, v >_{\tilde{H}} = < W, v >_{\tilde{H}} \qquad \forall\, v \in \tilde{H};$$

$$(ii) \lim_{n\to\infty} \|W_{1n} - W\|^2_{\tilde{L}^2_\varrho} = 0;$$

$$(iii) \lim_{n\to\infty} \|W_{1n1}\|_{\tilde{H}} + \|W_{1n3}\|_{\tilde{H}} = 0;$$

$$(iv)\ \lim_{n\to\infty} W_{1n}(x,t) = W(x,t) \quad \text{a.e. in } \tilde{\Omega}.$$

Now from (ii) above we see that

$$\hat{W}^c(j,k) = \lim_{n\to\infty} < W_{1n}, \tilde{\varphi}^c_{jk} >_{\tilde{\varrho}} \tag{6.17}$$

But for $1\le j \le j_o - 1$, $< W_{1n}, \tilde{\varphi}^c_{jk} >_{\tilde{\varrho}} = < W_{1n1}, \tilde{\varphi}^c_{jk} >_{\tilde{\varrho}}$. However from (iii) above and Schwarz's inequality, it follows that

$$\lim_{n\to\infty} < W_{1n1}, \tilde{\varphi}^c_{jk} >_{\tilde{\varrho}} = 0.$$

Consequently, we obtain from (6.17) that $\hat{W}^c(j,k) = 0$ for $1\le j \le j_o - 1$ and $\forall k$. Similar reasoning applies to $\hat{W}^c(j,k)$ for $j_o + j_1 \le j$, and also to $\hat{W}^s(j,k)$. We conclude

$$\hat{W}^c(j,k) = \hat{W}^s(j,k) = 0 \text{ for } 1 \le j \le j_o - 1,\ j_o + j_1 \le j, \text{ and } \forall k. \tag{6.18}$$

Next, using (6.7), we write

$$u_{1n2} = u^*_{1n2} + u^{**}_{1n2} \tag{6.19}$$

where

$$u^*_{1n2} = \sum_{j=j_o}^{j_o+j_1-1} \hat{u}^c_{1n}(j,0)\tilde{\varphi}^c_{j0} \quad \text{and}$$

$$u^{**}_{1n2} = \sum_{j=j_o}^{j_o+j_1-1} \sum_{k=1}^{n} [\hat{u}^c_{1n}(j,k)\tilde{\varphi}^c_{jk} + \hat{u}^s_{1n}(j,k)\tilde{\varphi}^s_{jk}].$$

We observe that $D_t u^*_{1n2} = 0$, and therefore that

$$D_t u^{**}_{1n2} = D_t u_{1n2}$$

and furthermore that

$$\|D_t u_{1n2}\|^2_{\tilde{L}^2_\varrho} \le \|D_t u_{1n}\|^2_{\tilde{L}^2_\varrho}.$$

Consequently, we have from (i) in (6.8) that

$$\lim_{n\to\infty} \|D_t u^{**}_{1n2}\|_{\tilde{L}^2_\varrho} / \|u_{1n}\|_{\tilde{L}^2_\varrho} = 0.$$

But an easy computation based on (6.19) shows that

$$\|u^{**}_{1n2}\|^2_{\tilde{L}^2_\varrho} \le \|D_t u^{**}_{1n2}\|^2_{\tilde{L}^2_\varrho}.$$

So we conclude from the last limit above that

$$\lim_{n\to\infty} \|u^{**}_{1n2}\|_{\tilde{L}^2_\varrho} / \|u_{1n}\|_{\tilde{L}^2_\varrho} = 0. \tag{6.20}$$

Next, using (6.15) and (6.19), we set

$$W^{*}_{1n2} = \frac{u^{*}_{1n2}}{\|u_{1n}\|_{\tilde{L}^2_\varrho}} \text{ and } W^{**}_{1n2} = \frac{u^{**}_{1n2}}{\|u_{1n}\|_{\tilde{L}^2_\varrho}}$$

and observe that

$$W_{1n2} = W^{*}_{1n2} + W^{**}_{1n2}.$$

Now it follows from (6.17) that for (j,k) fixed with $j_o \leq j \leq j_o + j_1 - 1$ and $1 \leq k$ that

$$\hat{W}^c(j,k) = \lim_{n\to\infty} < W^{**}_{1n2}, \tilde{\varphi}^c_{jk} >_{\tilde{\varrho}},$$

and consequently from (6.20) that for $j_o \leq j \leq j_o + j_1 - 1$ and $1 \leq k$

$$\hat{W}^c(j,k) = 0.$$

Similar reasoning shows that for $j_o \leq j \leq j_o + j_1 - 1$ and $1 \leq k$

$$\hat{W}^s(j,k) = 0$$

From these last two facts joined with (6.17)-(6.19), we obtain

$$(i)\, W(x,t) = \sum_{j=j_o}^{j_o+j_1-1} \hat{W}^c(j,0)\tilde{\varphi}^c_{j0} \text{ for (x,t)} \in \tilde{\Omega}, \tag{6.21}$$

$$(ii) \lim_{n\to\infty} W^{*}_{1n2}(x,t) = W(x,t) \text{ for (x,t)} \in \tilde{\Omega},$$

$$(iii) \lim_{n\to\infty} \|W^{*}_{1n2} - W\|^2_{\tilde{L}^2_\varrho} = 0.$$

We next replace v in (6.1) when i=1 by W^{*}_{1n2} to obtain

$$\mu_1 \mathcal{Q}^{\tilde{}}_1(u_{1n}, W^{*}_{1n2}) - \mu_1(\lambda_{j_o} + \frac{1}{n}) < u_{1n}, W^{*}_{1n2} >_{\tilde{\varrho}} = \tag{6.22}$$

$$< f_1(\cdot,\cdot,u_{1n}) + g_1(\cdot,\cdot,u_{1n},\ldots,u_{N_1+N_2+N_3 n}) - h_1(\cdot,\cdot), W^{*}_{1n2} >_{\tilde{\varrho}}$$

where we have used the fact that

$$< D_t u_{1n}, W^{*}_{1n2} >_{\tilde{\varrho}} = < D_t u^{**}_{1n2}, \frac{u^{*}_{1n2}}{\|u_{1n}\|_{\tilde{L}^2_\varrho}} >_{\tilde{\varrho}} = 0.$$

Now it follows from the definition of $P^o_{j_o}$ given in (ii) of (2.5) joined with (6.7) and (6.19) that

$$u^*_{1n2} = P^o_{j_o}(u_{1n}), \; W^*_{1n2} = P^o_{j_o}(u_{1n}) / \|u_{1n}\|_{\tilde{L}^2_\varrho}, \text{ and}$$

$$u_{1n} - P^o_{j_o}(u_{1n}) = u^{**}_{1n2} + u_{1n1} + u_{1n3}$$

Also,$\lambda_{j_o} \|u^{**}_{1n2}\|^2_{\tilde{L}^2_\varrho} = \|u^{**}_{1n2}\|^2_{\mathcal{L}^{\tilde{}}}$. We therefore obtain from (iii) of (6.8) and (6.20) that

$$\lim_{n\to\infty} \frac{\left\|u_{1n} - P^o_{j_o}(u_{1n})\right\|_{\tilde{L}^2_\varrho} + \left\|u_{1n} - P^o_{j_o}(u_{1n})\right\|_{\mathcal{L}^{\tilde{}}}}{\|u_{1n}\|_{\tilde{L}^2_\varrho} + \|u_{1n}\|_{\mathcal{L}^{\tilde{}}}} = 0$$

However λ_{j_0} is an $\tilde{\text{H}}$-L-pseudo-eigenvalue of $\tilde{\text{Q}}$. So we conclude from this last limit, (6.10), and (2.5) that

$$\lim_{n\to\infty} [\mathcal{Q}^{\tilde{}}(u_{1n}, P^o_{j_o}u_{1n}) - \mathcal{L}^{\tilde{}}(u_{1n}, P^o_{j_o}u_{1n})] / \|u_{1n}\|_{\tilde{L}^2_\varrho} = 0.$$

But this in turn implies by the above that

$$\lim_{n\to\infty} [\mathcal{Q}^{\tilde{}}(u_{1n}, W^*_{1n2}) - \mathcal{L}^{\tilde{}}(u_{1n}, W^*_{1n2})] = 0. \tag{6.23}$$

Next, we observe from (6.19) and the definition of W^*_{1n2} that

$$\mathcal{L}^{\tilde{}}(u_{1n}, W^*_{1n2}) = \lambda_{j_o} < u_{1n}, W^*_{1n2} >^{\tilde{}}_\varrho, \quad \text{and} \tag{6.24}$$

$$< u_{1n}, W^*_{1n2} >^{\tilde{}}_\varrho = \|u^*_{1n2}\|^2_{\tilde{L}^2_\varrho} / \|u_{1n}\|_{\tilde{L}^2_\varrho}.$$

Consequently, we have from (6.22)-(6.24) that

$$\limsup_{n\to\infty} < h_1(\cdot,\cdot), W^*_{1n2} >^{\tilde{}}_\varrho \geq \limsup_{n\to\infty} [< f_1(\cdot,\cdot,u_{1n}), W^*_{1n2} >^{\tilde{}}_\varrho$$

$$+ < g_1(\cdot,\cdot,u_{1n},\ldots,u_{N_1+N_2+N_3 n}), W^*_{1n2} >^{\tilde{}}_\varrho]. \tag{6.25}$$

Examining the expression involving g_1 on the right-hand of this last inequality, we see from (g-4) and (6.21)(iii) that

$$\lim_{n\to\infty} < g_1(\cdot,\cdot,u_{1n},\ldots,u_{N_1+N_2+N_3 n}), W^*_{1n2} >^{\tilde{}}_\varrho =$$

$$\lim_{n\to\infty} [\int_{\tilde{\Omega}\cap[W>0]} g_1(x,t,u_{1n},\ldots,u_{N_1+N_2+N_3 n}) W\varrho + \tag{6.26}$$

$$\int_{\tilde{\Omega}\cap[W<0]} g_1(x,t,u_{1n},\ldots,u_{N_1+N_2+N_3n})W\varrho].$$

But $u_{1n}(x,t) = \|u_{1n}\|_{\tilde{L}^2_\varrho} W_{1n}(x,t)$ and by (6.3) and (6.6), $\|u_{1n}\|_{\tilde{L}^2_\varrho} \to \infty$. Hence from (6.16)(iv) $u_{1n}(x,t) \to \infty$ a.e. on $\tilde{\Omega}\bigcap[W>0]$, and consequently by (g-4) and (g-5),

$$\lim_{n\to\infty} \int_{\tilde{\Omega}\cap[W>0]} g_1(x,t,u_{1n},\ldots,u_{N_1+N_2+N_3n})W\varrho = 0.$$

Likewise,

$$\lim_{n\to\infty} \int_{\tilde{\Omega}\cap[W<0]} g_1(x,t,u_{1n},\ldots,u_{N_1+N_2+N_3n})W\varrho = 0.$$

Therefore we obtain from (6.26) that

$$\lim_{n\to\infty} < g_1(\cdot,\cdot,u_{1n},\ldots,u_{N_1+N_2+N_3n}), W^*_{1n2} >_{\tilde{\varrho}} = 0.$$

Now $h_1 \in L^2_\varrho\left(\tilde{\Omega}\right)$. So from (6.21)(iii) we have that

$$\lim_{n\to\infty} < h_1(\cdot,\cdot), W^*_{1n2} >_{\tilde{\varrho}} = \int_{\tilde{\Omega}} h_1(x,t)W\varrho,$$

and we conclude from these last two limits joined with (6.25) that

$$\int_{\tilde{\Omega}} h_1(x,t)W\varrho \geq \limsup_{n\to\infty} < f_1(\cdot,\cdot,u_{1n}), W^*_{1n2} >_{\tilde{\varrho}}$$

Next, we observe from (f-10) and (6.21)(iii) that

$$\lim_{n\to\infty} < f_1(\cdot,\cdot,u_{1n}), W^*_{1n2} - W >_{\tilde{\varrho}} = 0$$

Consequently, from the last inequality above, we obtain that

$$\int_{\tilde{\Omega}} h_1(x,t)W\varrho \geq \limsup_{n\to\infty} [\int_{\tilde{\Omega}\cap[W>0]} f_1(x,t,u_{1n})W\varrho$$

$$+ \int_{\tilde{\Omega}\cap[W<0]} f_1(x,t,u_{1n})W\varrho] \tag{6.27}$$

Now as we have previously observed above,

$$u_{1n}(x,t) \to \infty \text{ a.e. on } \tilde{\Omega}\bigcap[W>0].$$

Therefore,

$$\liminf_{n\to\infty} f_1(x,t,u_{1n}) \geq f_{1+}(x,t) \qquad \text{a.e. in } \tilde{\Omega}\bigcap[W>0]$$

where $f_{1+}(x,t)$ is defined two lines below (f-10). Also,

$$f_1(x,t,u_{1n}) \geq -c(x,t) \qquad \text{a.e. in } \tilde{\Omega}\bigcap[W>0]$$

where $c(x,t) \in L^2_\varrho\left(\tilde{\Omega}\right)$. Therefore by Fatou's Lemma [Rud, p. 24], we have that

$$\liminf_{n\to\infty} \int_{\tilde{\Omega}\cap[W>0]} f_1(x,t,u_{1n})W\varrho \geq \int_{\tilde{\Omega}\cap[W>0]} f_{1+}(x,t)W\varrho.$$

In a similar manner, we obtain

$$\liminf_{n\to\infty} \int_{\tilde{\Omega}\cap[W<0]} f_1(x,t,u_{1n})W\varrho \geq \int_{\tilde{\Omega}\cap[W<0]} f_1^-(x,t)W\varrho.$$

We conclude from these last two inequalities in conjunction with (6.27) that

$$\int_{\tilde{\Omega}} h_1(x,t)W\varrho \geq \int_{\tilde{\Omega}\cap[W>0]} f_{1+}(x,t)W\varrho + \int_{\tilde{\Omega}\cap[W<0]} f_1^-(x,t)W\varrho. \qquad (6.28)$$

From (6.21)(i), we see that W(x) in the above inequality is a $\lambda_{j_o}-$eigenfunction of L where $\lambda_{j_o} = \lambda_{j_o(1)}$. Since $\|W_{1n}\|_{\tilde{L}^2_\varrho} = 1$, we see, furthermore, from (6.16)(ii) that $\|W\|_{\tilde{L}^2_\varrho} = 1$. Hence, W(x) is a nontrivial $\lambda_{j_o(1)}-$eigenfunction of L. But then the inequality in (6.28) is a direct contradiction of the inequality in (2.8) for i=1. We thus have that (6.3) has led us to contradiction.We conclude therefore that (6.2) is indeed correct for Case I, i.e., i=1,...,N_1.

We next establish (6.2) for Case II. Once again we will show that (6.2) holds for a particular i in Case II, namely i=N_1+1. A similar procedure will work for i=N_1+2,...,N_1+N_2. Also, we shall suppose $j_o(N_1+1)>1$. For $j_o(N_1+1)=1$ a similar proof will prevail. Suppose then that (6.2) is false for the sequence

$$\left\{\|u_{N_1+1n}\|_{\tilde{L}^2_\varrho} + \|u_{N_1+1n}\|_{\mathcal{L}^-}\right\}_{n=n_o^*+1}^{\infty}$$

Then without loss in generality, we can suppose that

$$\lim_{n\to\infty}[\|u_{N_1+1n}\|_{\tilde{L}^2_\varrho} + \|u_{N_1+1n}\|_{\mathcal{L}^-}] = \infty. \qquad (6.29)$$

The first part of the proof that (6.29) leads to a contradiction is identical with Case I. In particular,using (6.1) we obtain

$$< D_t u_{N_1+1n}, v >_{\tilde{\varrho}} + \mu_{N_1+1} \tilde{\mathcal{Q}}_{N_1+1}(u_{N_1+1n}, v) = \tag{6.30}$$

$$\mu_{N_1+1}(\lambda_{j_o(N_1+1)} + \varepsilon_{N_1+1}) < u_{N_1+1n}, v >_{\tilde{\varrho}} + < f_{N_1+1}(\cdot,\cdot,u_{N_1+1n})$$

$$+ g_{N_1+1}(\cdot,\cdot,u_{1n},\ldots,u_{N_1+N_2+N_3n}) - h_{N_1+1}(\cdot,\cdot), v >_{\tilde{\varrho}} \quad \forall v \in S_n^{\ddagger},$$

$$\text{where } 0 < \varepsilon_{N_1+1} < (\lambda_{j_o(N_1+1)+j_1(N_1+1)} - \lambda_{j_o(N_1+1)}),$$

and then following exactly the same procedure as in Case I, we arrive at the following four facts:

$$\|u_{N_1+1n}\|_{\mathcal{L}^{\cdot}} \leq K_2 \|u_{N_1+1n}\|_{\tilde{L}^2_{\varrho}} \quad \forall n > n_o^*; \tag{6..31}$$

$$(i) \lim_{n\to\infty} \|D_t u_{N_1+1n}\|_{\tilde{L}^2_{\varrho}} / \|u_{N_1+1n}\|_{\tilde{L}^2_{\varrho}} = 0; \tag{6.32}$$

$$(ii) \lim_{n\to\infty} [\|u_{N_1+1n1}\|^2_{\tilde{L}^2_{\varrho}} + \|u_{N_1+1n3}\|^2_{\mathcal{L}^{\cdot}}] / \|u_{N_1+1n}\|^2_{\tilde{L}^2_{\varrho}} = 0;$$

$$(iii) \lim_{n\to\infty} [\|u_{N_1+1n1}\|^2_{\tilde{H}} + \|u_{N_1+1n3}\|^2_{\tilde{H}}] / \|u_{N_1+1n}\|^2_{\tilde{L}^2_{\varrho}} = 0;$$

where

$$u_{N_1+1n} = u_{N_1+1n1} + u_{N_1+1n2} + u_{N_1+1n3} \tag{6.33}$$

where $j_o = j_o(N_1+1)$ and $j_1 = j_1(N_1+1)$ and

$$u_{N_1+1n1} = \sum_{j=1}^{j_o-1} \hat{u}^c_{N_1+1n}(j,0)\tilde{\varphi}^c_{j0}$$

$$+ \sum_{j=1}^{j_o-1}\sum_{k=1}^{n} [\hat{u}^c_{N_1+1n}(j,k)\tilde{\varphi}^c_{jk} + \hat{u}^s_{N_1+1n}(j,k)\tilde{\varphi}^s_{jk}],$$

$$u_{N_1+1n2} = \sum_{j=j_o}^{j_o+j_1-1} \hat{u}^c_{N_1+1n}(j,0)\tilde{\varphi}^c_{j0}$$

$$+ \sum_{j=j_o}^{j_o+j_1-1}\sum_{k=1}^{n} [\hat{u}^c_{N_1+1n}(j,k)\tilde{\varphi}^c_{jk} + \hat{u}^s_{N_1+1n}(j,k)\tilde{\varphi}^s_{jk}]$$

$$u_{N_1+1n3} = \sum_{j=j_o+j_1}^{n} \hat{u}^c_{N_1+1n}(j,0)\tilde{\varphi}^c_{j0}$$

$$+ \sum_{j=j_o+j_1}^{n} \sum_{k=1}^{n} [\hat{u}^c_{N_1+1n}(j,k)\tilde{\varphi}^c_{jk} + \hat{u}^s_{N_1+1n}(j,k)\tilde{\varphi}^s_{jk}].$$

Next, we observe that

$$\mathcal{L}^{\tilde{}}\,(u_{N_1+1n}, u_{N_1+1n2}) = \lambda_{j_o} < u_{N_1+1n}, u_{N_1+1n2} >^{\tilde{}}_{\varrho},$$

and from (2.3)(ii),(6.29), (6.31), and (6.32)(ii) that there is a positive constant K_3 such that

$$\left|\mathcal{Q}^{\tilde{}}\,(u_{N_1+1n}, u_{N_1+1n2}) - \mathcal{L}^{\tilde{}}\,(u_{N_1+1n}, u_{N_1+1n2})\right| \leq K_3 \left\|u_{N_1+1n2}\right\|_{\tilde{L}^2_\varrho}.$$

On putting u_{N_1+1n2} for v in (6.30) and using these last facts in conjunction with

$$< D_t u_{N_1+1n}, u_{N_1+1n2} >^{\tilde{}}_{\varrho} = 0,$$

we obtain that

$$\mu_{N_1+1}\varepsilon_{N_1+1} \left\|u_{N_1+1n2}\right\|_{\tilde{L}^2_\varrho} \leq \mu_{N_1+1}K_3 + \|c\|_{\tilde{L}^2_\varrho} + \|c^*\|_{\tilde{L}^2_\varrho} + \|h_{N_1+1}\|_{\tilde{L}^2_\varrho}$$

where we have also made use of (f-10) and (g-4).

Now it is clear from (6.29) and (6.31)-(6.33) that

$$\|u_{N_1+1n2}\|_{\tilde{L}^2_\varrho} \to \infty \quad \text{as n} \to \infty.$$

Hence on dividing both sides of this last inequality by $\|u_{N_1+1n2}\|_{\tilde{L}^2_\varrho}$ and leaving n$\to \infty$, we obtain that

$$\mu_{N_1+1}\varepsilon_{N_1+1} \leq 0.$$

But by the hypothesis of the theorem, both μ_{N_1+1} and ε_{N_1+1} are strictly positive quantities.. So we have arrived at a contradiction. Therefore (6.29) is false, and (6.2) is indeed true for i=N_1+1, and hence for i=$N_1+2, \ldots, N_1+N_2$ since the exact same proof just given will work for each of these values. The proof of (6.2) is therefore complete for Case II.

We next establish (6.2) for Case III. Once again we will show (6.2) holds for a particular value of i in Case III, namely i=N_1+N_2+1. In the ensuing discussion, we will designate this value by $\bar{N}$, i.e., $\bar{N}=N_1+N_2+1$. Suppose then that (6.2) is false for the sequence

$$\left\{\|u_{\bar{N}n}\|_{\tilde{L}^2_\varrho} + \|u_{\bar{N}n}\|_{\mathcal{L}^{\tilde{}}}\right\}_{n=n_o^*+1}^{\infty}.$$

where $u_{\bar{N}n} \in S_n^{\ddagger}$. Then without loss in generality we can suppose that

$$\lim_{n\to\infty} [\|u_{\bar{N}n}\|_{\tilde{L}^2_\varrho} + \|u_{\bar{N}n}\|_{\mathcal{L}^{\tilde{}}}] = \infty \tag{6.34}$$

We put $u_{\bar{N}n}$ in place of v in (6.1) when i=N̄ and obtain

$$\mu_{\bar{N}} \mathcal{Q}^{\tilde{}}_{\bar{N}} (u_{\bar{N}n}, u_{\bar{N}n}) - \mu_{\bar{N}}(\lambda_1 - \varepsilon_{\bar{N}}) < u_{\bar{N}n}, u_{\bar{N}n} >^{\tilde{}}_\varrho = \tag{6.35}$$

$$< f_{\bar{N}} (\cdot,\cdot, u_{\bar{N}n}) + g_{\bar{N}}(\cdot,\cdot, u_{1n}, \ldots, u_{N_1+N_2+N_3n}) - h_{\bar{N}}(\cdot,\cdot), u_{\bar{N}n} >^{\tilde{}}_\varrho$$

for n$\geq n_o^*+1$ where we have made use of the fact that $< u_{\bar{N}n}, D_t u_{\bar{N}n} >^{\tilde{}}_\varrho = 0$. It follows from (f-10) ,(g-4) and this last equality that

$$\mathcal{Q}^{\tilde{}}_{\bar{N}} (u_{\bar{N}n}, u_{\bar{N}n}) \leq \lambda_1 \|u_{\bar{N}n}\|^2_{\tilde{L}^2_\varrho}$$

$$+\mu_{\bar{N}}^{-1} \|u_{\bar{N}n}\|_{\tilde{L}^2_\varrho} [\|c\|_{\tilde{L}^2_\varrho} + \|c^*\|_{\tilde{L}^2_\varrho} + \|h_{\bar{N}}\|_{\tilde{L}^2_\varrho}]$$

where we also have made use of the fact that $\varepsilon_{\bar{N}}>0$. Now by (2.3)(ii)

$$\|u_{\bar{N}n}\|^2_{\mathcal{L}^{\tilde{}}} - \mathrm{K}_2 - K_1 \|u_{\bar{N}n}\|_{\mathcal{L}^{\tilde{}}} \leq \mathcal{Q}^{\tilde{}}_{\bar{N}} (u_{\bar{N}n}, u_{\bar{N}n})$$

for n$\geq n_o^* + 1$. Therefore, we conclude from these last two inequalities in conjunction with (6.34) that first

$$\lim_{n\to\infty} \|u_{\bar{N}n}\|_{\tilde{L}^2_\varrho} = \infty \tag{6.36}$$

and secondly that there exists a constant K_3 such that

$$\|u_{\bar{N}n}\|_{\mathcal{L}^{\tilde{}}} \leq K_3 \|u_{\bar{N}n}\|_{\tilde{L}^2_\varrho} \tag{6.37}$$

for n$\geq n_o^* + 1$.

Writing

$$u_{\bar{N}n} = \sum_{j=1}^n \hat{u}^c_{\bar{N}n}(j,0)\tilde{\varphi}^c_{j0} + \sum_{j=1}^n \sum_{k=1}^n [\hat{u}^c_{\bar{N}n}(j,k)\tilde{\varphi}^c_{jk} + \hat{u}^s_{\bar{N}n}(j,k)\tilde{\varphi}^s_{jk}],$$

we observe that

$$\mathcal{L}^{\tilde{}} (u_{\bar{N}n}, u_{\bar{N}n}) - \lambda_1 < u_{\bar{N}n}, u_{\bar{N}n} >^{\tilde{}}_\varrho =$$

$$\sum_{j=1}^n (\lambda_j - \lambda_1) \left|\hat{u}^c_{\bar{N}n}(j,0)\right|^2 + \sum_{j=1}^n (\lambda_j - \lambda_1) \sum_{k=1}^n [\left|\hat{u}^c_{\bar{N}n}(j,k)\right|^2 + \left|\hat{u}^s_{\bar{N}n}(j,k)\right|^2].$$

Hence,

$$\mathcal{L}^{\tilde{}}(u_{\bar{N}n}, u_{\bar{N}n}) - \lambda_1 < u_{\bar{N}n}, u_{\bar{N}n} >_{\varrho}^{\tilde{}} \geq 0.$$

Consequently, we see from (6.35) that

$$\mu_{\bar{N}}[\mathcal{Q}_{\bar{N}}^{\tilde{}}(u_{\bar{N}n}, u_{\bar{N}n}) - \mathcal{L}^{\tilde{}}(u_{\bar{N}n}, u_{\bar{N}n})] + \mu_{\bar{N}}\varepsilon_{\bar{N}} < u_{\bar{N}n}, u_{\bar{N}n} >_{\varrho}^{\tilde{}}$$

$$\leq < f_{\bar{N}}(\cdot,\cdot,u_{\bar{N}n}) + g_{\bar{N}}(\cdot,\cdot,u_{1n},\ldots,u_{N_1+N_2+N_3n}) - h_{\bar{N}}(\cdot,\cdot), u_{\bar{N}n} >_{\varrho}^{\tilde{}}$$

But then we obtain from (2.3)(ii), (f-10), (g-4) in conjunction with this last inequality that

$$\mu_{\bar{N}}\varepsilon_{\bar{N}} \|u_{\bar{N}n}\|^2_{\tilde{L}^2_{\varrho}} \leq \mu_{\bar{N}}[K_1 \|u_{\bar{N}n}\|_{\mathcal{L}^{\tilde{}}} + K_2]$$

$$+ \|u_{\bar{N}n}\|_{\tilde{L}^2_{\varrho}} [\|c\|_{\tilde{L}^2_{\varrho}} + \|c^*\|_{\tilde{L}^2_{\varrho}} + \|h_{\bar{N}}\|_{\tilde{L}^2_{\varrho}}]$$

Dividing both sides of this last inequality by $\|u_{\bar{N}n}\|^2_{\tilde{L}^2_{\varrho}}$ and passing to the limit as n$\rightarrow\infty$,and furthermore making use of (6.36) and (6.37), we obtain that

$$\mu_{\bar{N}}\varepsilon_{\bar{N}} \leq 0.$$

However, $\mu_{\bar{N}}$ and $\varepsilon_{\bar{N}}$ are strictly positive quantities, and we have arrived at a contradiction. Consequently, (6.34) is false, and (6.2) is indeed true for i $=\bar{N}$, i.e., i=N_1+N_2 + 1. A similar proof will work for the values of i=N_1+N_2 + 2,...,N_1+N_2+N_3. Hence, (6.2) is true for Case III, and we have succeeded in establishing (6.2) for all values of i, i.e., i=1,...,N_1+N_2+N_3.

Next, we show that there exists a constant K_1' such that

$$\|D_t u_{in}\|_{\tilde{L}^2_{\varrho}} \leq K_1' \qquad \forall n > n_0^* \quad \text{and i=1,...,} N_1 + N_2 + N_3. \tag{6.38}$$

We shall show that (6.38) holds in particular for u_{1n}. A similar proof will prevail for the other values of i.

To show that (6.38) holds for u_{1n}, we observe (as we have before) that $D_t u_{1n} \in S_n^{\ddagger}$, and then we replace v in (6.1) when i=1 with $D_t u_{1n}$ and obtain

$$\|D_t u_{in}\|^2_{\tilde{L}^2_{\varrho}} = < f_1(\cdot,\cdot,u_{1n}) + g_1(\cdot,\cdot,u_{1n},\ldots,u_{N_1+N_2+N_3n}) - h_1(\cdot,\cdot), D_t u_{1n} >$$

for n$\geq n_0^* + 1$ where we have used (Q-6) with Remark 2 and (6.5). From (f-10) and (g-4) in conjunction with this last equality, it follows that

$$\|D_t u_{in}\|^2_{\tilde{L}^2_{\varrho}} \leq [\|c\|_{\tilde{L}^2_{\varrho}} + \|c^*\|_{\tilde{L}^2_{\varrho}} + \|h_1\|_{\tilde{L}^2_{\varrho}}] \|D_t u_{in}\|_{\tilde{L}^2_{\varrho}}.$$

Hence,

$$\|D_t u_{in}\|_{\tilde{L}^2_\varrho} \leq [\|c\|_{\tilde{L}^2_\varrho} + \|c^*\|_{\tilde{L}^2_\varrho} + \|h_1\|_{\tilde{L}^2_\varrho}] \text{ for } n \geq n_0^* + 1,$$

and (6.38) is established for i=1. A similar proof works for all the other values of i, and, therefore, (6.38) is indeed true for i=1,...,$N_1+N_2+N_3$.

Next, we use (6.2) in conjunction with (6.38) and (1.2) and obtain that there exists a constant K^* such that

$$\|u_{in}\|_{\tilde{H}} \leq K^* \text{ for i=1,...,} N_1+N_2+N_3 \text{ and } n \geq n_0^*+1. \tag{6.39}$$

With (6.1) and this last inequality at our disposal, we proceed with the proof of the theorem along the lines of the proof of Theorem 1. In particular, we see from Lemma 3 and the observation that $\tilde{H}$ is a separable Hilbert space, that the following facts obtain for each of the sequences $\{u_{in}\}_{n=n_0^*+1}^{\infty}$ which satisfies (6.1) and (6.39) above for i=1,...,$N_1+N_2+N_3$:

$\exists\ u_i^* \in \tilde{H}$ and a subsequence (which for ease of notation we take to be the full sequence) such that

$$\lim_{n\to\infty} \|u_{in} - u_i^*\|_{\tilde{L}^2_\varrho} = 0, \tag{6.40}$$

$$\exists w \in \tilde{L}^2_\varrho \ s.t. \ |u_{in}(x,t)| \leq w(x,t) \tag{6.41}$$

$$\text{for a.e. } (x,t) \in \tilde{\Omega}, \quad n = n_o^*+1,\ n_o^*+2,\dots,$$

$$\lim_{n\to\infty} u_{in}(x,t) = u_i^*(x,t) \text{ for } a.e.\ (x,t) \in \tilde{\Omega}, \tag{6.42}$$

$$\lim_{n\to\infty} < D_k u_{in}, v >_{\tilde{p_k}} = < D_k u_i^*, v >_{\tilde{p_k}} \forall v \in \tilde{L}^2_{p_k} \text{ and } k = 1,...,N, \tag{6.43}$$

$$\lim_{n\to\infty} < D_t u_{in}, v >_{\tilde{\varrho}} = < D_t u_i^*, v >_{\tilde{\varrho}} \forall v \in \tilde{L}^2_\varrho, \tag{6.44}$$

for i=1,...,$N_1+N_2+N_3$.

Next, we propose to show the following: there exist subsequences

$$\{u_{1n_j}\}_{j=1}^{\infty}, \dots, \{u_{N_1+N_2+N_3 n_j}\}_{j=1}^{\infty}$$

such that

$$\lim_{j\to\infty} D_k u_{in_j}(x,t) = D_k u_i^*(x,t) \quad \text{for } a.e.\ (x,t) \in \tilde{\Omega} \tag{6.45}$$

for k=1,...,N and i=1,...,$N_1+N_2+N_3$.

We shall establish (6.45) for the particular case i=1. A very similar proof will work for i=2,...,$N_1+N_2+N_3$.

To show that (6.45) holds for the case i=1, we establish the following two facts:

(1) there exists a subsequence $\{u_{1n_j}\}_{j=1}^{\infty}$ such that

$$\lim_{j\to\infty}\sum_{k=1}^{N} p_k^{\frac{1}{2}}(x)\{[A_k(x,u_{1n_j}(x,t),Du_{1n_j}(x,t)) - A_k(x,u_{1n_j}(x,t), \tag{6.46}$$

$$Du_1^*(x,t))]\left[D_k u_{1n_j}(x,t) - D_k u_1^*(x,t)\right]\} = 0 \text{ for } a.e.\ (x,t) \in \tilde{\Omega}.$$

(2) With $\{u_{1n_j}\}_{j=1}^{\infty}$ designating the same subsequence as in (6.46),

$$\{| D_k u_{1n_j}(x,t) |\}_{j=1}^{\infty} \text{ is pointwise bounded for } a.e.\ (x,t) \in \tilde{\Omega} \tag{6.47}$$

for k=1,...,N.

The proof that (6.46) and (6.47) together imply (6.45) is very similar to the proof of the analogous situation given in Chapter 1 where we show (1,5.26) and (1,5.27) together imply (1,5.25). So we leave the details of the above implication, namely that (6.46) and (6.47) together imply (6.45), to the reader.

It remains to show that (6.46) and (6.47) hold.

To establish (6.46), we proceed along the same lines that we used to show that (5.60) in the proof of Theorem 1 holds. An examination of this proof shows that the main ingredient in the establishing of (6.46) is to show that the following fact holds:

$$\lim_{n\to\infty} \tilde{\mathcal{Q}_1}(u_{1n}, u_{1n} - u_1^*) = 0,$$

A further examination of the proof shows that this limit will hold provided we can show that

$$\lim_{n\to\infty} \tilde{\mathcal{Q}_1}(u_{1n}, u_{1n} - \tau_n(u_1^*)) = 0 \tag{6.48}$$

where $\tau_n(u_1^*)$ is defined by (4.8). So the proof of (6.46) will be complete once we show that using (6.1) and (6.39)-(6.44), it follows that (6.48) is indeed true. We now do this.

Since $\tau_n(u_1^*) \in S_n^{\ddagger}$, we see from (6.1) with i=1 that

$$< D_t u_{1n}, u_{1n} - \tau_n(u_1^*) >_{\tilde{\varrho}} + \mu_1 \tilde{\mathcal{Q}_1}(u_{1n}, u_{1n} - \tau_n(u_1^*))$$

$$= \mu_1(\lambda_{j_o(1)} + n^{-1}) < u_{1n}, u_{1n} - \tau_n(u_1^*) >_{\tilde{\varrho}} \tag{6.49}$$

$$+ < f_1(\cdot,\cdot,u_{1n}) + g_1(\cdot,\cdot,u_{1n},...,u_{N_1+N_2+N_3 n}) - h_1(\cdot,\cdot), u_{1n} - \tau_n(u_1^*) >_{\tilde{\varrho}}.$$

Next, we observe from (6.40) that

$$\lim_{n\to\infty} \|u_{1n} - u_1^*\|_{\tilde{L}^2_\varrho} = 0,$$

and from Lemma 1 that

$$\lim_{n\to\infty} \|u_1^* - \tau_n(u_1^*)\|_{\tilde{L}^2_\varrho} = 0.$$

Hence,

$$\lim_{n\to\infty} \|u_{1n} - \tau_n(u_1^*)\|_{\tilde{L}^2_\varrho} = 0, \tag{6.50}$$

and consequently we see from (6.39) and (6.49) that

$$\lim_{n\to\infty} \mu_1 \tilde{\mathcal{Q}_1}(u_{1n}, u_{1n} - \tau_n(u_1^*)) =$$

$$\lim_{n\to\infty} < f_1(\cdot,\cdot,u_{1n}) + g_1(\cdot,\cdot,u_{1n},...,u_{N_1+N_2+N_3 n}) - h_1(\cdot,\cdot), u_{1n} - \tau_n(u_1^*) >_{\tilde{\varrho}}$$

But by (f-10), $\|f_1(\cdot,\cdot,u_{1n})\|_{\tilde{L}^2_\varrho} \le \|c\|_{\tilde{L}^2_\varrho}$, by (g-4),

$$\|g_1(\cdot,\cdot,u_{1n},...,u_{N_1+N_2+N_3 n})\|_{\tilde{L}^2_\varrho} \le \|c^*\|_{\tilde{L}^2_\varrho},$$

and also $\|h_1\|_{\tilde{L}^2_\varrho}$ is finite. So we obtain from (6.50) and Schwarz's inequality that the limit on the right-hand side of this last equation is zero. Hence , the limit on the left is zero. On the other hand, μ_1 is a strictly positive quantity. Consequently,

$$\lim_{n\to\infty} \tilde{\mathcal{Q}_1}(u_{1n}, u_{1n} - \tau_n(u_1^*)) = 0,$$

and the proof of (6.48) is complete. As we mentioned earlier, this shows that (6.46) is indeed true for i=1. A similar proof prevails for the other values of i. So (6.46) holds for i=1,...,$N_1+N_2+N_3$.

Next, we establish (6.47). We do this once again for the case i=1. A similar procedure works for i=2,...,$N_1+N_2+N_3$. To establish (6.47) for i=1, we first observe that

$$|Du(x,t)| = \left[|D_1 u(x,t)|^2 + \ldots + |D_n u(x,t)|^2\right]^{\frac{1}{2}}.$$

Next, we let $\tilde{\Omega}_2$ be the set where simultaneously

$$u_1^*(x,t), |Du_1^*(x,t)|, \ h_k^*(x), \ u_{1n_j}(x,t), \ A_k\left(x, u_{1n_j}(x), Du_{1n_j}(x,t)\right)$$

$$\text{and } A_k\left(x, u_{1n_j}(x,t), u_1^*(x,t)\right) \text{ are finite-valued}$$

for k=1,...,N and j=1,2,...,where (Q-2) and (Q-3) hold, and also where the limits in (6.46) and (6.42) for i=1 exist. Then

$$meas\ (\tilde{\Omega} - \tilde{\Omega}_2) = 0. \tag{6.51}$$

Suppose, to the contrary, that $\{| Du_{1n_j}(x,t) |\}_{j=1}^{\infty}$ is not a pointwise bounded sequence for each $(x,t) \in \tilde{\Omega}_2$. Then there exists $(x_o, t_o) \in \tilde{\Omega}_2$ and a subsequence $\left\{| Du_{1n_{j_l}}(x_o, t_o) |\right\}$ such that

$$\lim_{l\to\infty} \left| Du_{1n_{j_l}}(x_o, t_o) \right| = \infty \tag{6.52}$$

Set

$$c_3 = min\left[p_1(x_o), ..., p_N(x_o)\right]. \text{ Hence, } c_3 > 0. \tag{6.53}$$

From (Q-3), we have

$$\sum_{k=1}^{N} p_k^{\frac{1}{2}}(x_o) A_k\left(x_o, u_{1n_{j_l}}(x_o,t_o), Du_{1n_{j_l}}(x_o,t_o)\right) D_k u_{1n_{j_l}}(x_o,t_o) \tag{6.54}$$

$$\geq c_2 c_3 \left| Du_{n_{j_l}}(x_o,t_o)\right|^2.$$

Also,

$$\sum_{k=1}^{N} p_k^{\frac{1}{2}}(x_o) A_k\left(x_o, u_{1n_{j_l}}(x_o,t_o), Du_{n_{j_l}}(x_o,t_o)\right) D_k u_{1n_{j_l}}(x_o,t_o) \tag{6.55}$$

$$= \sum_{k=1}^{N} p_k^{\frac{1}{2}}(x_o) A_k\left(x_o, u_{1n_{j_l}}(x_o,t_o), D_k u_{1n_{j_l}}(x_o,t_o)\right) D_k u_1^*(x_o,t_o)$$

$$+ \sum_{k=1}^{N} p_k^{\frac{1}{2}}(x_o) A_k\left(x_o, u_{1n_{j_l}}(x_o,t_o), Du_1^*(x_o,t_o)\right) [D_k u_{1n_{j_l}}(x_o,t_o) -$$

$$D_k u_1^*(x_o,t_o)] + \sum_{k=1}^{N} p_k^{\frac{1}{2}}(x_o) [A_k\left(x_o, u_{1n_{j_l}}(x_o,t_o), Du_{1n_{j_l}}(x_o,t_o)\right) -$$

$$A_k\left(x_o, u_{1n_{j_l}}(x_o,t_o), Du_1^*(x_o,t_o)\right)][D_k u_{1n_{j_l}}(x_o,t_o) - D_k u_1^*(x_o,t_o)].$$

We divide both sides of (6.54) by $\left|Du_{1n_{j_l}}(x_o,t_o)\right|^{3/2}$, pass to the limit as $l\to\infty$, and conclude from (6.55), (Q-2), (6.42), (6.46), and the definition of $\tilde{\Omega}_2$ that

$$0\geq c_2c_3\lim_{l\to\infty}\left|Du_{1n_{j_l}}(x_o,t_o)\right|^{\frac{1}{2}}.$$

From (Q-3) and (6.53), we furthermore observe that $c_2c_3>0$. But this last inequality then implies that

$$\lim_{l\to\infty}\left|Du_{1n_{j_l}}(x_o,t_o)\right|=0.$$

However, this last limit is a direct contradiction of (6.52). Consequently, the sequence $\left\{\left|Du_{1n_j}(x,t)\right|\right\}_{j=1}^{\infty}$ is pointwise bounded at every $(x,t)\in\tilde{\Omega}_2$. This fact in conjunction with (6.51) establishes (6.47). As we mentioned earlier, (6.47) along with (6.46) gives (6.45) for i=1. An exactly similar proof works to show that (6.45) holds for the other values of i, namely i=2,...,$N_1+N_2+N_3$. So (6.45) is completely established.

Proceeding with the proof of Theorem 2, we now have that (6.39)-(6.45) hold and also that (6.1) holds. What we want to show is that (2.9) holds. Once again, we shall give the proof of (2.9) for the particular case i=1. A similar proof will work for the other values of i.

So with $i=1$, let $v\in\tilde{H}$. Then $\tau_J(v)\in S_J^{\ddagger}$ where $J\geq n_o^*+1$ and $\tau_J(v)$ is defined by (4.8). Then we have from (6.1) that

$$\begin{aligned}&<D_tu_{1n},\tau_J(v)>_{\tilde{\varrho}}+\mu_1\tilde{\mathcal{Q}}_1(u_{1n},\tau_J(v))= \qquad (6.56)\\ &\mu_1(\lambda_{j_o(1)}+\frac{1}{n})<u_{1n},\tau_J(v)>_{\tilde{\varrho}}+<f_1(\cdot,\cdot,u_{1n})\\ &+g_1(\cdot,\cdot,u_{1n},\ldots,u_{N_1+N_2+N_3n})-h_1(\cdot,\cdot),\tau_J(v)>_{\tilde{\varrho}}\end{aligned}$$

for $n\geq J$.

Now just as in the proof at this point of Theorem 1, we have from (6.39)-(6.45) that

$$\lim_{n\to\infty}\tilde{\mathcal{Q}}_1(u_{1n},\tau_J(v))=\tilde{\mathcal{Q}}_1(u_1^*,\tau_J(v)).$$

Likewise we have from (f-9), (f-10), (g-3), (g-4), and (6.42) that

$$\begin{aligned}&\lim_{n\to\infty}<f_1(\cdot,\cdot,u_{1n})+g_1(\cdot,\cdot,u_{1n},...,u_{N_1+N_2+N_3n}),\tau_J(v)>_{\tilde{\varrho}}=\\ &<f(\cdot,\cdot,u_1^*)+g_1(\cdot,\cdot,u_1^*,...,u_{N_1+N_2+N_3n}^*),\tau_J(v)>_{\tilde{\varrho}}.\end{aligned}$$

So using (6.39)-(6.45) in conjunction with these last two limits, we obtain from (6.56) that

$$< D_t u_1^*, \tau_J(v) >_{\varrho}^{\sim} + \mu_1 \tilde{\mathcal{Q}}_1 (u_1^*, \tau_J(v)) = \qquad (6.57)$$

$$\mu_1(\lambda_{j_o(1)} + \frac{1}{n}) < u_1^*, \tau_J(v) >_{\varrho}^{\sim} + < f_1 (\cdot, \cdot, u_1^*)$$

$$+ g_1(\cdot, \cdot, u_1^*, \ldots, u_{N_1+N_2+N_3}^*) - h_1(\cdot, \cdot), \tau_J(v) >_{\varrho}^{\sim}.$$

From Lemma 1, we see that

$$\lim_{n \to \infty} \|\tau_n (v) - v\|_{\tilde{H}} = 0 \quad \forall v \in \tilde{H}.$$

Hence, using this limit in conjunction (6.57) enables us to conclude that

$$< D_t u_1^*, v >_{\varrho}^{\sim} + \mu_1 \tilde{\mathcal{Q}}_1 (u_1^*, v) =$$

$$\mu_1(\lambda_{j_o(1)} + \frac{1}{n}) < u_1^*, v >_{\varrho}^{\sim} + < f_1 (\cdot, \cdot, u_1^*)$$

$$+ g_1(\cdot, \cdot, u_1^*, \ldots, u_{N_1+N_2+N_3}^*) - h_1(\cdot, \cdot), v >_{\varrho}^{\sim}.$$

But this is exactly (2.9) for i=1. Similar reasoning using (6.1) and (6.39)-(6.45), gives us (2.9) for $i = 2, \ldots, N_1 + N_2 + N_3$, and the proof of Theorem 2 is complete.

2.7 References

[BC] R. L. Borrelli and C. S. Coleman, *Differential Equations, a Modeling Approach*, Prentice Hall, Englewood Cliffs, N. J., 1987.

[BD] W. E. Boyce and R. C. Diprima, *Elementary Differential Equations and Boundary Value Problems,* Sixth Edition, John Wiley & Sons, New York, 1997.

[BdF] H. Berestycki and D. G. de Figueredo, *Double Resonance in Semilinear Elliptic Problems, Comm. Partial Differential Equations,* **6**(1981), 91-120.

[Be] E. Beltrami, *Mathematics for Dynamic Modeling,* Academic Press, SanDiego, 1987.

[BJS] L. Bers, F. John, and M. Schechter, *Partial Differential Equations,* John Wiley, New York, 1966.

[BN] H. Brezis and L. Nirenberg, *Characterization of ranges of some nonlinear operators and applications of boundary value problems,* Ann. Scuo. Norm. Sup. Pisa **5** (1978), 225-326.

[BR] F. E. Browder, *Existence theorems in partial differential equations,* in Proc. Symposia in Pure Mathematics, vol. 16, pp. 1-60, Amer. Math. Soc., Providence, 1970.

[CH] R. Courant and D. Hilbert, *Methods of Matematical Physics,* vol. 1, John Wiley, New York, 1966.

[CL] A. Castro and A. C. Lazer, *Results on periodic solutions of parabolic equations suggested by elliptic theory,* Bollettino U.M.I. (6) **1**-B (1982), 1089-1104.

[dFG] D. G. de Figueredo and J. P. Gossez, *Nonlinear partial differential equations near the first eigenvalue,* J. Diff. Eqns. **30**(1978), 1-19.

[DS] N. Dunford and J. Schwartz, *Linear Operators Part I: General Theory,* Wiley-Interscience, New York, 1957.

[EK] L. Edelstein -Keshet, *Mathematical Models in Biology*, Random House, New York, 1988.

[Fi] P. C. Fife, *Mathematical Aspects of Reacting and Diffusing Systems,* Lecture Notes in Biomathematics, vol. 28, Springer-Verlag, Berlin Heidelberg New York,1979.

[GF] I. M. Gelfand and S. V. Fomin, *Calculus of Variations*, Prentice-Hall, Englewood Cliffs, 1963.

[GT] D. Gilbarg and N. S. Trudinger, *Elliptic Partial Differential Equations of Second Order,* 2nd Edition, Springer-Verlag, Berlin, 1983.

[Het] G. Hetzer, *On a Semilinear Operator Equation at Resonance*, Houston Journal of Mathematics, **6**(1980), 277-285.

[Hel] G, Helwig, *Partial Differential Equations*, Blaisdell, New York, 1964.

[Ho] H. Hochstadt, *The Functions of Mathematical Physics,* Wiley and Interscience, New York, 1971.

[Ke] S. Kesavan, *Topics in Functional Analysis and Applications,* John Wiley & Sons, New York, 1989.

[KS] D. Kinderlehrer and G. Stampacchia, *An Introduction to Variational Inequalities and Their Applications,* Academic Press, New York, 1980.

[Ku] A. Kufner, *Weighted Soblev Spaces,* 2nd Edition, John Wiley & Sons, New York, 1985.

[KW] J. L. Kazdan and F. W. Warner, *Remarks on some quasilinear elliptic equations,* Comm. Pure Appl. Math., **28**(1975), 567-597.

[La] A. C. Lazer, Book Review, *Solvability of nonlinear equations and boundary value problems,* Bull. Amer. Math.Soc., **8**(1983), 482-489.

[LfS] L. Lefton and V. L. Shapiro, *Resonance and quasilinear parabolic differential equations,* J. Diff. Eqns. **101**(1993), 148-177.

[LgS] M. Legner and V. L. Shapiro, *Time-periodic quasilinear reaction-diffusion equations,* SIAM J. Math Anal., **26**(1996), 135-169.

[LaL] E. M. Landesman and A. C. Lazer, *Nonlinear perturbations of linear elliptic boundary value problems,* J. Math. Mech., **7**(1970), 609-623.

[LeL] J. Leray and J. L. Lions, *Quelques résultats de Visik sur les problèmes elliptiques non linéaires par les méthodes de Minty-Browder,* Bull. Soc. Math. France, **93**(1965), 97-107.

[Mu1] J. D. Murray, *How the leopard gets its spots*, Scientific American, **258**(1988), 80-87.

[Mu2] J. D. Murray, *Mathematical Biology,* Springer-Verlag, New York, 1990.

[MW] P. J. McKenna and W. Walter, *On the multiplicity of the solution set of some nonlinear boundary value problems,* Nonlinear Analysis TMA, **8**(1984), 893-907.

[Ni] L. Nirenberg, *Topics in Nonlinear Functional Analysis,* Courant Institute of Mathematical Sciences, New York University, New York, 1974.

[Ra] E. D. Rainville, *Special Functions*, Macmillan, New York, 1960

[Rob] S. Robinson, Double resonance in semilinear elliptic boundary value problems over bounded and unbounded domains, Nonlinear Analysis TMA, **21**(1993), 407-424.

[Rud] W. Rudin, *Real and Complex Analysis*, Second ed., McGraw-Hill, New York, 1974.

[Rum] A. Rumbos, *A semilinear elliptic boundary value problem at resonance where the nonlinearity may grow linearly,* Nonlinear Analysis TMA, **16**(1991), 1159-1168.

[Sh1] V. L. Shapiro, *Special functions and singular quasilinear partial differential equations,* SIAM J. Math Anal., 22(1991), 1411-1429.

[Sh2] V. L. Shapiro, *Quasilinear ellipticity and the 1st eigenvalue*, Comm. Partial Differential Equations 16 (1991), 1819-1855.

[Sh3] V. L. Shapiro, *Resonance, distributions and semilinear elliptic partial differential equations,* Nonlinear Analysis TMA, **8**(1984), 857-871.

[Wi] S. Williams, *A sharp sufficient condition for a solution of a nonlinear boundary value problem,* J. Diff. Eqns. **8**(1970), 580-588.

[Zy] A. Zygmund, *Trigonometric Series* vol. I, Cambridge Univ. Press, Cambridge, 1959.

DEPARTMENT of MATHEMATICS, UNIVERSITY OF CALIFORNIA, RIVERSIDE, RIVERSIDE, CALIFORNIA 92521-0135

E-mail address: shapiro@math.ucr.edu

Editorial Information

To be published in the *Memoirs*, a paper must be correct, new, nontrivial, and significant. Further, it must be well written and of interest to a substantial number of mathematicians. Piecemeal results, such as an inconclusive step toward an unproved major theorem or a minor variation on a known result, are in general not acceptable for publication. Papers appearing in *Memoirs* are generally longer than those appearing in *Transactions*, which shares the same editorial committee.

As of May 31, 2001, the backlog for this journal was approximately 7 volumes. This estimate is the result of dividing the number of manuscripts for this journal in the Providence office that have not yet gone to the printer on the above date by the average number of monographs per volume over the previous twelve months, reduced by the number of volumes published in four months (the time necessary for preparing a volume for the printer). (There are 6 volumes per year, each containing at least 4 numbers.)

A Consent to Publish and Copyright Agreement is required before a paper will be published in the *Memoirs*. After a paper is accepted for publication, the Providence office will send a Consent to Publish and Copyright Agreement to all authors of the paper. By submitting a paper to the *Memoirs*, authors certify that the results have not been submitted to nor are they under consideration for publication by another journal, conference proceedings, or similar publication.

Information for Authors

Memoirs are printed from camera copy fully prepared by the author. This means that the finished book will look exactly like the copy submitted.

The paper must contain a *descriptive title* and an *abstract* that summarizes the article in language suitable for workers in the general field (algebra, analysis, etc.). The *descriptive title* should be short, but informative; useless or vague phrases such as "some remarks about" or "concerning" should be avoided. The *abstract* should be at least one complete sentence, and at most 300 words. Included with the footnotes to the paper should be the 2000 *Mathematics Subject Classification* representing the primary and secondary subjects of the article. The classifications are accessible from `www.ams.org/msc/`. The list of classifications is also available in print starting with the 1999 annual index of *Mathematical Reviews*. The Mathematics Subject Classification footnote may be followed by a list of *key words and phrases* describing the subject matter of the article and taken from it. Journal abbreviations used in bibliographies are listed in the latest *Mathematical Reviews* annual index. The series abbreviations are also accessible from `www.ams.org/publications/`. To help in preparing and verifying references, the AMS offers MR Lookup, a Reference Tool for Linking, at `www.ams.org/mrlookup/`. When the manuscript is submitted, authors should supply the editor with electronic addresses if available. These will be printed after the postal address at the end of the article.

Electronically prepared manuscripts. The AMS encourages electronically prepared manuscripts, with a strong preference for $\mathcal{A}_{\mathcal{M}}\mathcal{S}$-LaTeX. To this end, the Society has prepared $\mathcal{A}_{\mathcal{M}}\mathcal{S}$-LaTeX author packages for each AMS publication. Author packages include instructions for preparing electronic manuscripts, the *AMS Author Handbook*, samples, and a style file that generates the particular design specifications of that publication series. Though $\mathcal{A}_{\mathcal{M}}\mathcal{S}$-LaTeX is the highly preferred format of TeX, author packages are also available in $\mathcal{A}_{\mathcal{M}}\mathcal{S}$-TeX.

Authors may retrieve an author package from e-MATH starting from `www.ams.org/tex/` or via FTP to `ftp.ams.org` (login as `anonymous`, enter username as password, and type `cd pub/author-info`). The *AMS Author Handbook* and the *Instruction Manual* are available in PDF format following the author packages link from `www.ams.org/tex/`. The author package can be obtained free of charge by sending email to `pub@ams.org` (Internet) or from the Publication Division, American Mathematical Society, P.O. Box 6248, Providence, RI 02940-6248. When requesting an author package, please specify $\mathcal{A}\mathcal{M}\mathcal{S}$-LaTeX or $\mathcal{A}\mathcal{M}\mathcal{S}$-TeX, Macintosh or IBM (3.5) format, and the publication in which your paper will appear. Please be sure to include your complete mailing address.

Sending electronic files. After acceptance, the source file(s) should be sent to the Providence office (this includes any TeX source file, any graphics files, and the DVI or PostScript file).

Before sending the source file, be sure you have proofread your paper carefully. The files you send must be the EXACT files used to generate the proof copy that was accepted for publication. For all publications, authors are required to send a printed copy of their paper, which exactly matches the copy approved for publication, along with any graphics that will appear in the paper.

TeX files may be submitted by email, FTP, or on diskette. The DVI file(s) and PostScript files should be submitted only by FTP or on diskette unless they are encoded properly to submit through email. (DVI files are binary and PostScript files tend to be very large.)

Electronically prepared manuscripts can be sent via email to `pub-submit@ams.org` (Internet). The subject line of the message should include the publication code to identify it as a Memoir. TeX source files, DVI files, and PostScript files can be transferred over the Internet by FTP to the Internet node `e-math.ams.org` (130.44.1.100).

Electronic graphics. Comprehensive instructions on preparing graphics are available at `www.ams.org/jourhtml/graphics.html`. A few of the major requirements are given here.

Submit files for graphics as EPS (Encapsulated PostScript) files. This includes graphics originated via a graphics application as well as scanned photographs or other computer-generated images. If this is not possible, TIFF files are acceptable as long as they can be opened in Adobe Photoshop or Illustrator. No matter what method was used to produce the graphic, it is necessary to provide a paper copy to the AMS.

Authors using graphics packages for the creation of electronic art should also avoid the use of any lines thinner than 0.5 points in width. Many graphics packages allow the user to specify a "hairline" for a very thin line. Hairlines often look acceptable when proofed on a typical laser printer. However, when produced on a high-resolution laser imagesetter, hairlines become nearly invisible and will be lost entirely in the final printing process.

Screens should be set to values between 15% and 85%. Screens which fall outside of this range are too light or too dark to print correctly. Variations of screens within a graphic should be no less than 10%.

Inquiries. Any inquiries concerning a paper that has been accepted for publication should be sent directly to the Electronic Prepress Department, American Mathematical Society, P. O. Box 6248, Providence, RI 02940-6248.

Selected Titles in This Series

(Continued from the front of this publication)

696 **Donald J. Estep, Mats G. Larson, and Roy D. Williams,** Estimating the error of numerical solutions of systems of reaction-diffusion equations, 2000

695 **Vitaly Bergelson and Randall McCutcheon,** An ergodic IP polynomial Szemerédi theorem, 2000

694 **Alberto Bressan, Graziano Crasta, and Benedetto Piccoli,** Well-posedness of the Cauchy problem for $n \times n$ systems of conservation laws, 2000

693 **Doug Pickrell,** Invariant measures for unitary groups associated to Kac-Moody Lie algebras, 2000

692 **Mara D. Neusel,** Inverse invariant theory and Steenrod operations, 2000

691 **Bruce Hughes and Stratos Prassidis,** Control and relaxation over the circle, 2000

690 **Robert Rumely, Chi Fong Lau, and Robert Varley,** Existence of the sectional capacity, 2000

689 **M. A. Dickmann and F. Miraglia,** Special groups: Boolean-theoretic methods in the theory of quadratic forms, 2000

688 **Piotr Hajłasz and Pekka Koskela,** Sobolev met Poincaré, 2000

687 **Guy David and Stephen Semmes,** Uniform rectifiability and quasiminimizing sets of arbitrary codimension, 2000

686 **L. Gaunce Lewis, Jr.,** Splitting theorems for certain equivariant spectra, 2000

685 **Jean-Luc Joly, Guy Metivier, and Jeffrey Rauch,** Caustics for dissipative semilinear oscillations, 2000

684 **Harvey I. Blau, Bangteng Xu, Z. Arad, E. Fisman, V. Miloslavsky, and M. Muzychuk,** Homogeneous integral table algebras of degree three: A trilogy, 2000

683 **Serge Bouc,** Non-additive exact functors and tensor induction for Mackey functors, 2000

682 **Martin Majewski,** ational homotopical models and uniqueness, 2000

681 **David P. Blecher, Paul S. Muhly, and Vern I. Paulsen,** Categories of operator modules (Morita equivalence and projective modules, 2000

680 **Joachim Zacharias,** Continuous tensor products and Arveson's spectral C^*-algebras, 2000

679 **Y. A. Abramovich and A. K. Kitover,** Inverses of disjointness preserving operators, 2000

678 **Wilhelm Stannat,** The theory of generalized Dirichlet forms and its applications in analysis and stochastics, 1999

677 **Volodymyr V. Lyubashenko,** Squared Hopf algebras, 1999

676 **S. Strelitz,** Asymptotics for solutions of linear differential equations having turning points with applications, 1999

675 **Michael B. Marcus and Jay Rosen,** Renormalized self-intersection local times and Wick power chaos processes, 1999

674 **R. Lawther and D. M. Testerman,** A_1 subgroups of exceptional algebraic groups, 1999

673 **John Lott,** Diffeomorphisms and noncommutative analytic torsion, 1999

672 **Yael Karshon,** Periodic Hamiltonian flows on four dimensional manifolds, 1999

671 **Andrzej Rosłanowski and Saharon Shelah,** Norms on possibilities I: Forcing with trees and creatures, 1999

670 **Steve Jackson,** A computation of δ_5^1, 1999

669 **Seán Keel and James M^c^Kernan,** Rational curves on quasi-projective surfaces, 1999

668 **E. N. Dancer and P. Poláčik,** Realization of vector fields and dynamics of spatially homogeneous parabolic equations, 1999